D1068409

Eugene C. Ogden
Gilbert S. Raynor
Janet V. Hayes
Donald M. Lewis
John H. Haines

MANUAL FOR SAMPLING AIRBORNE POLLEN

MANUAL FOR SAMPLING AIRBORNE POLLEN

MANUAL FOR SAMPLING AIRBORNE POLLEN

Eugene C. Ogden
New York State Museum and Science Service

Gilbert S. Raynor
Brookhaven National Laboratory

Janet V. Hayes
New York State Museum and Science Service

Donald M. Lewis
New York State Museum and Science Service

John H. Haines
New York State Museum and Science Service

Prepared under the auspices of the New York State Museum and Science Service and the U.S. Atomic Energy Commission and supported, in part, by Research Grant No. R-800677, from the Division of Meteorology, U.S. Environmental Protection Agency.

HAFNER PRESS
A Division of Macmillan Publishing Co., Inc.
New York
Collier Macmillan Publishers
London

Copyright © 1974 by Hafner Press

All rights reserved. No part of this book may be reproduced or transmitted in any form or by any means, electronic or mechanical, including photocopying, recording, or by any information storage and retrieval system, without permission in writing from the Publisher.

Hafner Press

A Division of Macmillan Publishing Co., Inc.
866 Third Avenue, New York, N.Y. 10022
Collier-Macmillan Canada Ltd.

Library of Congress Cataloging in Publication Data

Main entry under title:

Manual for sampling airborne pollen.

Prepared under the auspices of the New York State Museum and Science Service and the U.S. Atomic Energy Commission.

1. Allergens—Measurement. 2. Pollen. 3. Air—Analysis. I. Ogden, Eugene Cecil. II. New York (State). State Museum and Science Service. III. United States. Atomic Energy Commission.

RA577.P6M36 614.7'1 74-4397

ISBN 0-02-849820-8

Printed in the United States of America

Cover Illustration

Ambrosia artemisiifolia pollen grain. Scanning electron micrograph at a magnification of 3,000 diameters. Photo by and courtesy of Eric Lifshin, General Electric Research and Development Center, Schenectady, New York.

RA
577
.P6M36

Dedicated
to the memory of
Oren C. Durham

YOUNGSTOWN STATE UNIVERSITY
LIBRARY

335393

Contents

Acknowledgments

Among the many persons who have given us much appreciated help and encouragement, a few warrant special mention:

Nancy Farr (B.S., Russell Sage College), experienced in sampling and identification of airborne pollen and proficient in grammar and punctuation, saw the manuscript through all of its typing and proofreading stages.

Eileen Coulston (M.S. in Plant Science, Syracuse University), with several years experience as a science librarian, served as our bibliographer. Her suggestions have added greatly to clarity in the text.

Dr. Donald L. Collins, Director of the Biological Survey, N.Y.S. Museum and Science Service, read the manuscript and suggested several improvements.

Richard Snook (B.A. in Botany, University of Miami) prepared the drawings for 17 full-page figures.

Gwyneth Gillette (B.A. in Art, Cornell College), prepared all of the drawings and charts for Chapter 9 and figure 1 and the chart for Chapter 10.

Dr. Allan Solomon, Department of Geosciences, University of Arizona, sent prints of several published and unpublished scanning electron microscope photographs of pollen grains. Dr. Paul Martin kindly gave permission to use them.

Research grants which have supported our studies in pollen sampling have come from the National Institute of Allergy and Infectious Diseases (E-1958), Division of Air Pollution, Bureau of State Services (AP-81), and the Division of Meteorology, Environmental Protection Agency (R-800677). To these agencies and to the New York State Museum and Science Service and to Brookhaven National Laboratory, we express our grateful thanks.

"Hurry up with that pollen count!"

Courtesy of Medical Tribune

1

INTRODUCTION

Sampling and identification of atmospheric pollen is, by its nature, a botanical endeavor. However, this Manual is written primarily for nonbotanists which requires the inclusion of much elementary botanical information. Users may find it desirable to supplement their knowledge by recourse to botanical publications. We have attempted to introduce the subject, outlining the elementary procedures and suggesting sources for further information. Some of the remarks here may seem rather elementary, but it should be kept in mind that to some persons such information may be new.

Determining the concentrations of pollen in the air may involve many related activities, depending upon the completeness and accuracy required. This book is deliberately elementary. It introduces the investigator to the subject in its various aspects:

1. Choice of the proper sampling instruments.
2. Selection of sites that adequately reflect the desired information.
3. Acquaintance with local plants that supply pollen to the atmosphere.
4. Methods for preparing the samples.
5. Identification of pollen and some other organic particles.
6. Data analysis and reporting.

Bibliographic references at the end of each Chapter suggest additional sources of information. They are not exhaustive. They are selected for their practical value. Recent publications with bibliographies are often given preference over earlier, historically more important, ones.

This Manual should find use among physicians, medical technicians, health officers, sanitary engineers, meteorologists, geochronologists, and geneticists; all those who have occasion to make "pollen counts," those whose activities include determinations of concentrations of pollen in the atmosphere, and those who would interpret the data reported by others.

AEROBIOLOGY

The study of pollen and the spores of such plants as fungi, algae, mosses, and ferns is palynology, a term coined by Hyde and Williams in 1944. It was defined as "the study of pollens and other plant spores and their dispersal, and applications thereof." Studies that deal primarily with those that are airborne may be called aeropalynology. They are a part of a larger field of investigation that includes animal as well as plant particles in the atmosphere. In practically all situations, the air contains many kinds of contaminants, organic and inorganic, having great diversity in size, shape, density, and many other characteristics. Studies on the biological constituents in contrast to the inorganic gases, liquids, and solid particles but in relation to the meteorological and other factors is the broad field of aerobiology.

This field of science includes: identification, morphology, physiology, viability, longevity, sampling, concentrations, diurnal and seasonal patterns, phenology, emission, transport, dispersion, pollination, pollinosis, temperature, lapse rate, wind speed and direction, turbulence, precipitation, and a host of other subjects. As Dr. Gregory words it: "Aerobiology is usually understood to be the study of passively airborne micro-organisms—of their identity, behaviour, movements, and survival."

KINDS OF AIRBORNE PARTICLES

Solid particles in the atmosphere may be organic or inorganic. Organic particles include pollen grains; spores of fungi, mosses, and ferns; plant hairs and other fragments; fungus hyphae; alga cells and filaments; animal dander; insect scales; and other miscellaneous materials. These are routinely found to be a part of the catch when sampling the air for particles in the size range of pollen. Inorganic materials commonly caught are: soil particles, tar, oil drops, and salt crystals.

Although this Manual is designed for guidance in sampling pollen in the atmosphere, it is judged advisable to include a chapter on the identification of those fungus spores that appear in the samples. Sampling is here defined to include both catching and identifying the particles of interest.

POLLEN COUNTS

Pollen counts vary with the purpose and the available facilities. They may be merely qualitative: indicators of presence or absence of certain pollens with only a general idea as to abundance. Preferably, they also give reasonably accurate quantitative indications: average number per unit volume. In some situations, only a single kind of pollen has interest and is the only one counted and reported, as during the ragweed season. In North America, at least during August and September, when the "pollen count" is reported via the news media without specifying the kind of pollen, it is the pollen of various species of ragweed (*Ambrosia*) that is meant. Other pollens that occur in the samples are not included. At other times, during the spring tree pollen season especially, several kinds of pollen (pine, oak, birch, and a score of others) may be recognized, counted, and recorded. Frequently, some of the larger, recognizable fungus spores are included.

Spot samples

Average pollen concentrations during short periods of time, such as a few minutes, are usually difficult to obtain with desirable accuracy due to the small amount of air that is involved and the consequent low number of pollen grains. At least one cubic meter of air should be included and preferably several. For this reason, periods of one or two hours can be considered to be spot samples. Such samples have value in research studies for correlating with other data: weather conditions, medical observations, pollination times, or simultaneous spot samples at other locations. Spot or short-period samples taken consecutively indicate diurnal patterns. In most situations, a single spot sample is unreliable for indicating the daily average concentration.

Daily samples

Much of the sampling of airborne pollen is on a 24-hour basis. The type of sampler that is used determines if the sampling surface is exposed continuously or intermittently to obtain an adequate sample without overloading to a degree that decreases accuracy. The reported count is the average during the period, expressed as the average number of pollen grains per unit volume of the air sampled and ideally is representative of concentrations in the vicinity of the sampler. The time of beginning and ending the day's sample is usually chosen with consideration for the operator, so is generally at some daylight hour. Thus, the daily sample would not represent a discreet calendar day (i.e., midnight to midnight). For many of the purposes using daily pollen counts this is of little importance. In situations requiring data for calendar days, sequential samplers may be employed and serviced at any time of the day as the data are easily grouped into any desired time periods. With some types of samplers, a pair may be operated side by side with an automatic timer set to stop one and activate the other.

Sequential sampling

Consecutive short-period samples yield data in sequence which are more meaningful than the average for a long period. The long-period averages, when desired, may be computed easily from the sequential record. Several kinds of samplers, employing different principles of sampling techniques, yield sequential data. Some which are suitable for particles in the pollen size range are discussed in Chapter 6.

Fluctuations in the weather, especially rain and wind direction, may cause large variations in pollen concentrations from hour to hour. The diurnal patterns of pollen emission from the sources may affect the patterns of concentration in some locations. In many localities, especially those distant from sources of the pollens of interest where the concentrations are strongly dependent on the vagaries of weather, these concentrations can vary greatly over relatively short periods of time. As an example, in the Adirondack Mountains of New York State, ragweed pollen may vary from zero to several hundred grains per cubic meter during a day. The day's average often does not give a clue to a heavy concentration that persists for a few hours.

POLLEN SURVEYS

The ideal pollen survey combines field observations, with carefully made quantitative observations over a period of several years, of the seasonal patterns of those pollens which occur in significant concentrations. A large number of surveys have been made with varying degrees of accuracy and detail. A report that lumps all kinds of pollen into a single category is essentially worthless. Even a grouping into trees, grasses, and weeds has little value. The more useful surveys separate the various genera.

Unfortunately, the pollens of some closely related plants are difficult or impossible to distinguish, which will require that they be lumped together in the survey reports unless field observations on the presence, location, abundance, and time of pollen shed of the suspected source plants yield useful clues. Generally, in routine surveys of atmospheric pollen, it is not easily possible to separate species within a genus and sometimes not even genera within a family. The use for which the survey is made will dictate, to some extent, the accuracy that is required. Allergists, plant breeders, plant geographers, and paleobotanists may have somewhat different requirements for the survey reports to be useful to them.

References

Gregory, P.H. 1973. The microbiology of the atmosphere. 2nd rev. ed. John Wiley & Sons, New York. 377 p.

Gregory, P.H., and J.L. Monteith, eds. 1967. Airborne microbes. Cambridge Univ. Press, London. 385 p.

Moulton, F.R., ed. 1942. Aerobiology. AAAS [Amer. Ass. Advan. Sci., Washington, D.C.] Publ. 17. 289 p.

Raynor, G.S., E.C. Ogden, D.M. Lewis, and J.V. Hayes. Aerobiology. *Submitted for* M.B. Rhyne, ed. Childhood hay fever, to be published by C.C. Thomas, Springfield, Ill.

2
RECOGNITION OF PLANTS

Identification of atmospheric pollen and the conduct of pollen surveys are greatly facilitated by a knowledge of the local flora, especially of those plants that are anemophilous (pollen carried primarily on air currents). Where they grow in relation to the sampler, their abundance, when they pollinate, and their habits with regard to pollen production and dispersal can properly influence analysis of the samples. Ignoring such information can cause embarrassing errors and erroneous reports.

Although pollen is usually identified to genus only, it is desirable to recognize species in the local flora. Some genera are composed of several species which vary greatly in the quantity of pollen that becomes airborne. Also, species of a genus may have different pollination times.

SOURCES OF INFORMATION

Every area of temperate North America is covered by some sort of publication on the flowering plants and conifers of the area. These vary from simple lists to local and regional floristic treatments to detailed descriptive and illustrated manuals. In each area, the number of species that produce pollen that is carried by air currents in sizable amounts an appreciable distance from the source is not large.

For the nonbotanist, popular books of the "how-to-know" type with illustrations are best. Those with line drawings, especially if details of leaves, flowers, and fruits are shown, make recognition of common plants relatively easy. Most publications of this sort do not distinguish the plants on the basis of the agency that transports the pollen. From a phylogenetic standpoint this is not as fundamental as many other characteristics, nor is it especially useful in learning to recognize one plant from another.

Plants that are anemophilous may not always be readily distinguished from those that are entomophilous (pollen carried primarily by insects). However, during anthesis (flowering time) one can get a rather good idea by tapping a flower or cluster of flowers onto the hand. If the pollen leaves the flowers freely and appears dry, it is likely that the mode of travel from flower to flower or plant to plant is at least partly by air. In most cases it will be necessary to identify the plants without regard to this character, after which further inquiry in the literature or elsewhere will determine its mode of pollination.

Botanists in the area, both professional and amateur, are usually willing to assist. Today, nearly all parts of the world are within visiting distance of institutions of higher learning which include attention to plant science (botany). Visits to botany departments of local colleges and consultations with staff members, especially taxonomists, ecologists, and herbarium curators, are recommended. Those establishments that maintain plant collections, either preserved or growing in gardens, are especially helpful.

METHODS OF STUDY

Classification

Plants that produce pollen are the seed plants which include the angiosperms (flowering plants) and the gymnosperms (conifers and related groups). They are grouped into classes, orders, families, genera, and species. For each plant, it is important to know the genus, species, and family to which it belongs.

Closely related genera (based on morphology and genetics) will be grouped into one family. The similarity may extend to the pollen which might be nearly indistinguishable and may have similar allergenic properties. For example, some genera in the birch family (Betulaceae) such as birch (*Betula*), hop hornbeam (*Ostrya*), and blue beech (*Carpinus*) have pollen that is difficult to separate in routine counting.

Nomenclature

Plants do not have names in the sense that they have leaves and flowers. The names by which we know them are the names we have given them. Such names are for two purposes: for convenience and for showing relationships. All human languages have their own popular or vernacular names for the common plants, but in all areas many rare or inconspicuous kinds have no popular names at all. On the other hand, many of the very common, showy, and widely distributed plants have many different popular names.

Persons accustomed to using one local name for a plant may not recognize it by other local names. A common tree (*Larix laricina*) is variously known as larch, tamarack, and hackmatack. More unfortunately, the same name may be used in different parts of the country for different plants. Ironwood is used for both *Ostrya virginiana* (also know as hop hornbeam) and *Carpinus caroliniana* (also called American hornbeam, blue beech, water beech, musclewood, and other local names). Beech or American beech generally refers to *Fagus grandifolia.* Sycamore in North America means *Platanus occidentalis,* but in England it means *Acer pseudo-platanus.* To report pollen merely as being from cedar would mean different genera to readers in different areas: *Cedrus, Libocedrus, Thuja, Chamaecyparis, Juniperus,* or even *Lycopodium.* Several genera are called pigweed (chiefly *Chenopodium* and *Amaranthus*), so qualifying terms should be added. June grass may be a species of *Poa, Danthonia, Koeleria,* and perhaps of other genera. Gum might mean *Nyssa* (black gum, sour gum, tupelo, or pepperidge) or *Liquidambar* (sweet gum or red gum). In the usual sense, box elder is not an elder, sweet fern is not a fern, mountain ash is not an ash, blue beech is not a beech, red cedar is not a cedar, and ground hemlock is not a hemlock.

These examples, from plants with airborne pollen, serve to caution the investigator against too much reliance on common provincial names. In this Manual, which draws its examples primarily from North America, but which will find use in other parts of the world, the scientific names are used or accompany the commonly heard English names. Scientific names are international names and have greater stability. Each scientific name of a plant is composed of a genus (noun) and species (modifier). Subspecific categories may be part of the name but are unnecessary in pollen surveys. Author citations are for bibliographic purposes and usually may be ignored. Scientific names should be used for exchange of information among professionals, such as: botanists, meteorologists, and physicians. They may be obtained from any modern floristic publication.

Persons unacquainted with the principles of scientific botanical nomenclature may be confused when, consulting several floristic publications, they find that more than one scientific name is being used for the same kind of plant. It is not our purpose here to deal with the many reasons for this. Any textbook on plant taxonomy (also called systematic botany) will have one or more chapters devoted to the subject. However, a few examples may help to clarify the usual situations. The genus *Ambrosia* includes many species, most of them commonly known as "ragweed." One of these species is abundant along roadsides, in waste places, gardens, and cultivated fields in North America. Generally, it is known as *Ambrosia artemisiifolia*. This is the plant, in most areas of the United States, that is the principle cause of pollinosis or hayfever. As with many species of plants, it exhibits variation: in the dissection of its leaves, in the size of the flowers, and in the kinds or amounts of pubescence (plant hairs) on the stems and leaves. Some of these characteristics may be inherited (genetic), others may be adaptations to the habitat (ecological). Some botanists would prefer to treat as two or more species what other botanists consider to be but one. This is a matter of opinion and each opinion may be as justifiable as the other. The international rules of botanical nomenclature make no attempt to force an opinion, but after an opinion has been decided upon then the rules are to be followed in determining the correct or valid name. This means, however, that an individual plant may have more than one valid name, depending on one's opinion with regard to its similarity and differences in comparison with other closely related plants.

A. artemisiifolia is considered by some to include a wide range of variations in appearance. The variations might then be treated as subspecies, or varieties, or forms, or other categories, indicating that these differences are not as distinct as they should be for the separation of species. Others might prefer to recognize two species here. One of these would be *A. artemisiifolia* and would be the variant that agrees with the original description of that species. The other would take another species name, the one that was first proposed for it. Actually, a Swedish botanist in the middle 1700's, writing under the pen name of Carolus Linnaeus, was the first to name *A. artemisiifolia* to include the taxon (taxonomic category) having coarsely dissected leaves and large male flower clusters and *A. elatior* for the plant with more finely dissected leaves and smaller flower clusters. It then becomes a matter of opinion and our choice whether to call the commoner, more

YOUNGSTOWN STATE UNIVERSITY
LIBRARY
335393

widespread, finely dissected-leaved plant. *A. elatior* (if we think there are two separate species here) or *A. artemisiifolia* (if we include both variants in one species). If one wishes to recognize the differences at some lesser rank than species, he might refer to these taxa as *A. artemisiifolia* variety *artemisiifolia* and *A. artemisiifolia* var. *elatior,* as is commonly done. For those not experienced in such matters, it is well to follow some floristic manual in the choice of names.

Species of the genus *Ambrosia* are commonly known as ragweeds. A group of species, very closely related to those in *Ambrosia,* are in the genus *Franseria.* They are sometimes referred to as the false ragweeds. Recent studies indicate that *Franseria* is too closely allied to *Ambrosia* to be treated as a separate genus. Thus, one of the species may be listed as *Franseria acanthicarpa* or as *Ambrosia acanthicarpa.* Both names are valid and both refer to the same plant which is commonly known as bur ragweed because of its prickly fruits.

There are many other reasons for the use of more than one name for a plant, some valid, others invalid, but they need not be discussed here.

Habit

This refers to the nature of the species which includes: size; whether annual, biennial or perennial; whether herbaceous or woody; abundance of cones or flowers; quantity of pollen produced; and the size, shape, and density of its pollen. Some of these characteristics may influence the amount of pollen that is produced. Height above ground at which the pollen is shed greatly determines the distance to which pollen will travel in quantity.

Whether a plant is an annual, biennial, or herbaceous perennial may not always be easy to determine. This may be important from the standpoint of eradication but perhaps not with regard to quantity of pollen produced. All trees and shrubs are perennial. Biennial plants are scarcely found among those that have airborne pollen, but a few, such as sugar beet (*Beta vulgaris*) when grown for seed, may be a local source of atmospheric pollen.

With most conifers, the female (seed-bearing) cones are easily located, when present, but the male (pollen-bearing) cones may be inconspicuous. Unfortunately, few books dealing with identification of gymnosperms include illustrations of the staminate cones and many ignore the inconspicuous ones, even in the descriptions. Generally, close examination of the branches will determine if pollen-bearing structures are present.

The situation is much better with regard to published illustrations of the flowers that produce pollen, although the less conspicuous flowers of the plants whose pollen is airborne are not as frequently illustrated as are those that are large and showy.

Habitat

A knowledge of the type of locality or ecological conditions under which the plants grow will help to determine probable sources of the local atmospheric pollens. For example, species of *Ambrosia* are primarily plants of open disturbed soil rather than forests, hayfields, or marshes.

Distribution and abundance

It is important to know which anemophilous plants are widespread or abundant and which are local or scattered or rare; also their distance from the sampling station (or from hayfever patients) and direction with regard to the prevailing winds. An otherwise unexplained unusually high or low count of a particular pollen might be due to an air mass movement. A local source may produce concentrations of pollen in a small downwind region several orders of magnitude above the concentrations of this pollen from numerous more distant sources.

Phenology

This is the science of the relations of climate and periodic biological phenomena, such as the shedding of pollen. One of the useful aids in identification of pollen is the knowledge of pollination times in comparison to dates the samples were obtained. Frequent field observations should be made on the development of flowers. Annual records over a period of a few years will enable one to predict, for local areas, the onset of pollen from the different species. It is well to include in the phenology list some plants that have conspicuous flowers, even though their pollen is not airborne. Such records will greatly help in the comparisons of seasons from year to year. It will be found that some plants flower at approximately the same time each year, while others vary greatly.

References

Benson, L. 1957. Plant classification. D.C. Heath & Co., Boston. 688 p.

Core, E.L. 1955. Plant taxonomy. Prentice-Hall, Inc., Englewood Cliffs, N.J. 459 p.

Davis, P.H., and V.H. Heywood. 1963. Principles of angiosperm taxonomy. D. Van Nostrand, Princeton, N.J. 556 p.

Lawrence, G.H.M. 1951. Taxonomy of vascular plants. Macmillan Co., New York. 823 p.

Porter, C.L. 1967. Taxonomy of flowering plants. 2nd ed. W.H. Freeman & Co., San Francisco. 472 p.

Solomon, W.R., O.C. Durham, and F.L. McKay. 1967. Pollens and the plants that produce them. Pages 340–397 *in* J.M. Sheldon, R.G. Lovell, and K.P. Mathews, A manual of clinical allergy. 2nd ed. W.B. Saunders Co., Philadelphia.

Wodehouse, R.P. 1971. Hayfever plants. 2nd rev. ed. Hafner Publ. Co., New York. 280 p.

3 SOURCES OF AIRBORNE POLLEN

Pollen is produced by the seed plants which include the gymnosperms (conifers and related groups) and the angiosperms (flowering plants). Pollen grains are male reproductive structures of seed plants. They function to transport the male gametes (sperm) to the female gametes (eggs) where fertilization may occur. This transport is pollination, which must not be confused with fertilization (union of a sperm nucleus with an egg nucleus) which occurs later.

Conifers produce pollen in cones that are separate from those that produce the seeds (Figs. 3-1 and 3-2). With some species, both types of cones may be found on the same plant (monoecious). With others they are on separate plants (dioecious). Flowering plants may produce pollen in the same flowers that produce the seeds (bisexual or perfect), or in separate flowers but on the same plant (monoecious), or only on separate plants (dioecious). If both bisexual and unisexual flowers are on the same plant, it is called polygamous. Thus, it might be polygamo-monoecious or polygamo-dioecious.

Information as to which plants have perfect flowers and which plants are monoecious, dioecious, or combinations of these is incomplete in nearly all manuals. The following list indicates the usual situation for some of the genera that produce airborne pollen in North America.

COMMON GENERA OF AIRBORNE POLLENS

Ginkgoaceae

Ginkgo, ginkgo, maidenhair tree (dioecious)

Taxaceae

Taxus, yew (dioecious, rarely monoecious) Fig. 3-5A

Pinaceae

Pinus, pine (monoecious) Figs. 3-1 and 3-5C and D
Picea, spruce (monoecious)
Abies, fir (monoecious)

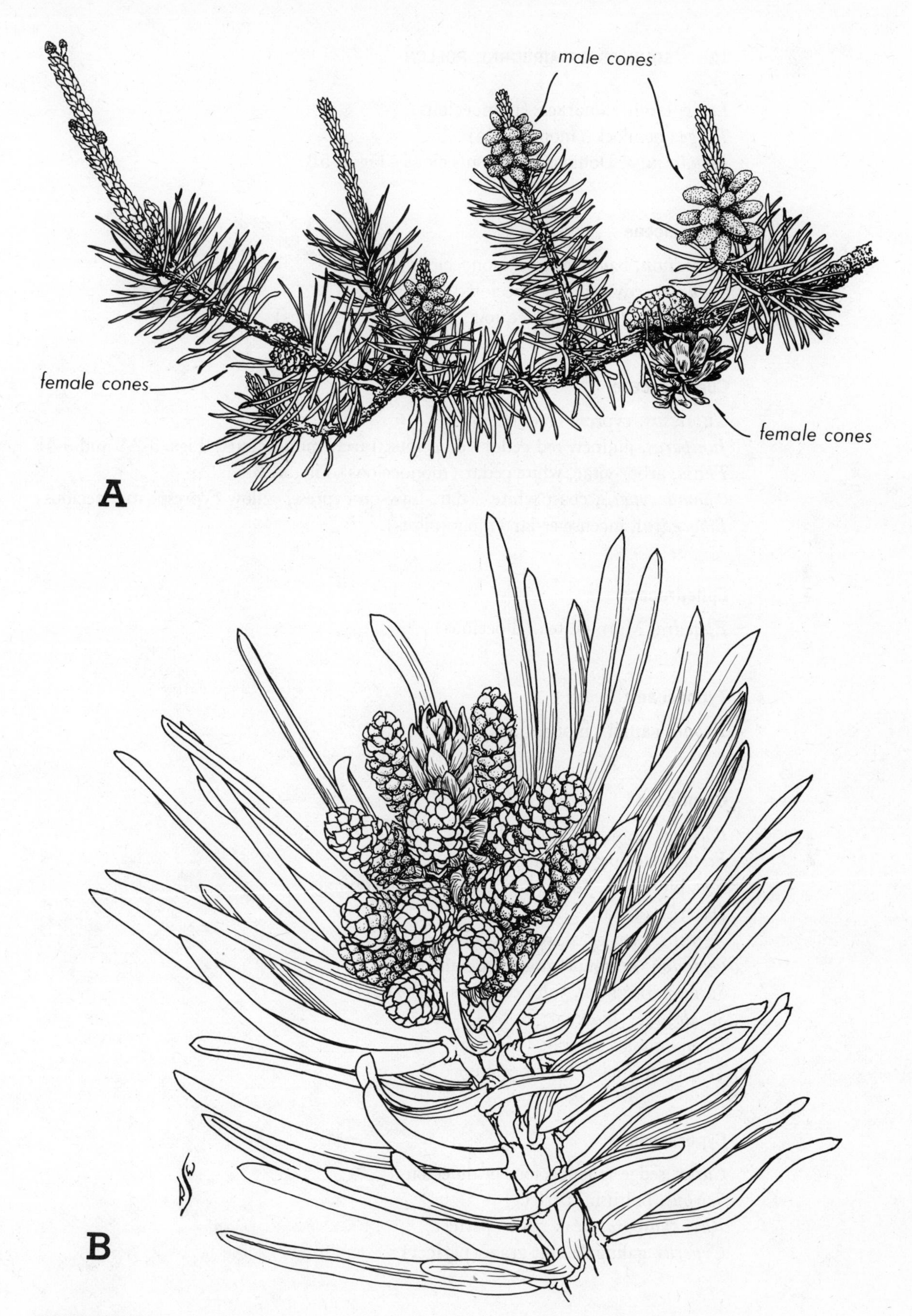

FIGURE 3-1 Pinus banksiana. A, branch with male and female cones. B, cluster of male cones.

Larix, larch, tamarack (monoecious)
Tsuga, hemlock (monoecious)
Pseudotsuga, Douglas fir (monoecious) Fig. 3-5B

Taxodiaceae

Taxodium, bald cypress (monoecious)
Sequoia, redwood (monoecious) Fig. 3-5E
Sequoiadendron, giant sequoia, bigtree (monoecious)

Cupressaceae

Cupressus, cypress (monoecious)
Juniperus, juniper, red cedar (dioecious, rarely monoecious) Figs. 3-2A and 3-5F
Thuja, arbor-vitae, white cedar (monoecious) Fig. 3-2B
Chamaecyparis, coast white cedar, Lawson cypress, yellow cypress (monoecious)
Libocedrus, incense cedar (monoecious)

Ephedraceae

Ephedra, Mormon tea (dioecious)

Typhaceae

Typha, cattail (monoecious)

Gramineae*

Phleum, timothy (bisexual)
Festuca, fescue (bisexual)
Poa, blue grass, June grass, spear grass (bisexual)
Dactylis, orchard grass (bisexual)
Agrostis, bent grass, redtop (bisexual)
Cynodon, Bermuda grass (bisexual)
Sorgum, sorghum, Johnson grass (bisexual)
Secale, rye (bisexual)
Phragmites, reed grass (bisexual)
Zea, corn, maize (monoecious)

Cyperaceae

Carex, sedge (monoecious or dioecious)
Scirpus, bulrush (perfect)
Rhynchospora, beak rush (perfect)
Cyperus, galingale, nut grass (perfect)

* A grass spikelet is a group of flowers, some of which may be perfect and others unisexual. If both sexes occur in the same spikelet, it is here indicated to be bisexual rather than polygamous.

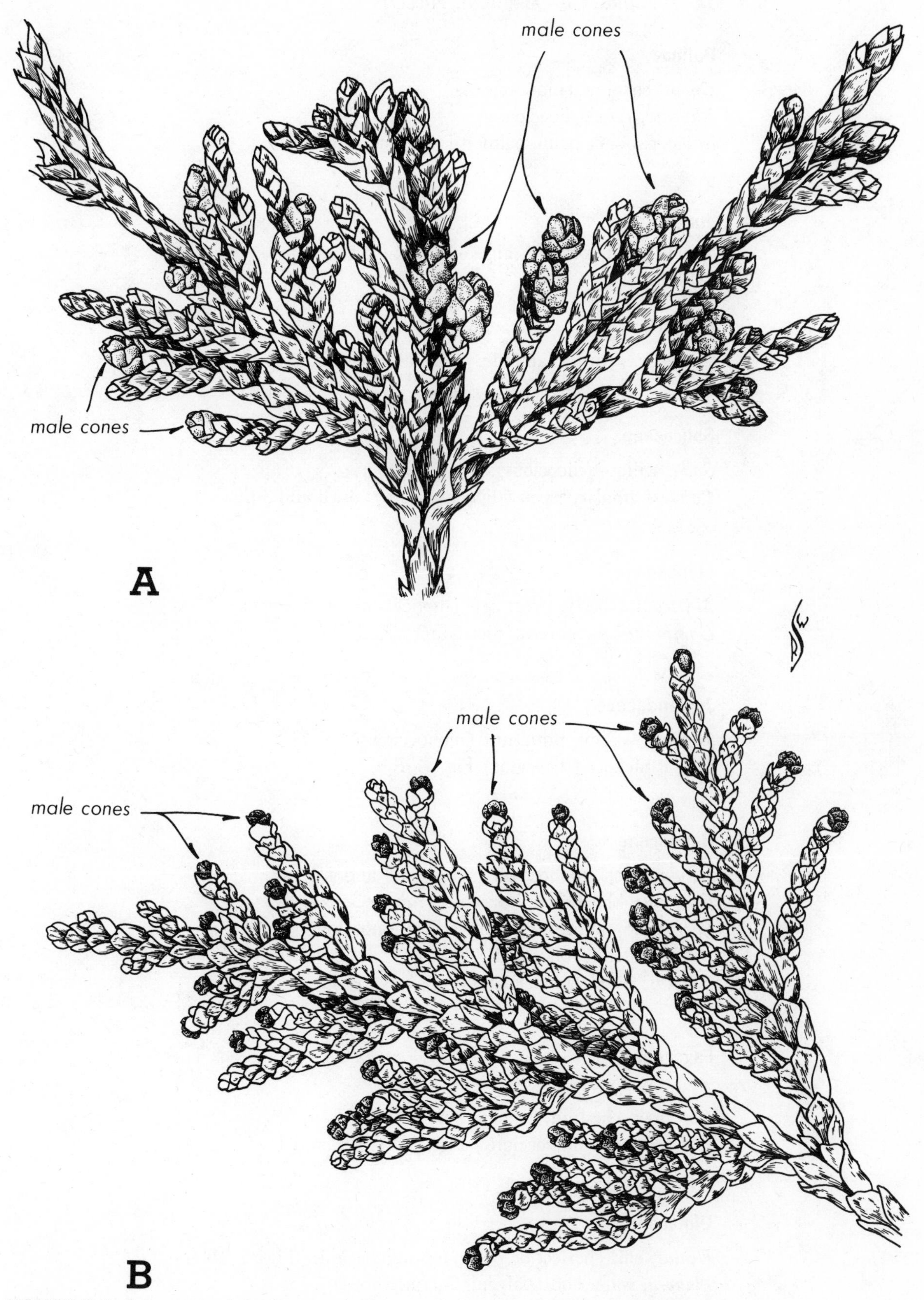

FIGURE 3-2 A, branch of Juniperus virginiana. B, branch of Thuja occidentalis.

Palmae

Cocos, coconut (monoecious)
Phoenix, date (dioecious)
Sabal, cabbage palm, palmetto (perfect) Fig. 3-6B

Juncaceae

Juncus, rush (perfect) Fig. 3-4A

Casuarinaceae

Casuarina, Australian pine, joint fir (monoecious or dioecious)

Salicaceae

Salix, willow (dioecious)
Populus, poplar, aspen (dioecious) Figs. 3-4B and 3-6C

Myricaceae

Myrica, bayberry, sweet gale (monoecious or dioecious)
Comptonia, sweet fern (dioecious) Fig. 3-6D

Juglandaceae

Juglans, walnut, butternut (monoecious)
Carya, hickory (dioecious) Fig. 3-6E

Betulaceae

Betula, birch (monoecious) Figs. 3-6F and 4-4
Carpinus, blue beech, ironwood (monoecious)
Ostrya, hop hornbeam, ironwood (monoecious)
Alnus, alder (monoecious)
Corylus, hazel (monoecious)

Fagaceae

Fagus, beech (monoecious) Fig. 4-6
Castanea, chestnut (monoecious)
Quercus, oak (monoecious) Figs. 3-3A, 4-5, and 4-7

Ulmaceae

Ulmus, elm (perfect or polygamo-monoecious) Figs. 3-3B and 3-7A
Planera, water elm (polygamo-monoecious)
Celtis, hackberry (polygamo-monoecious)

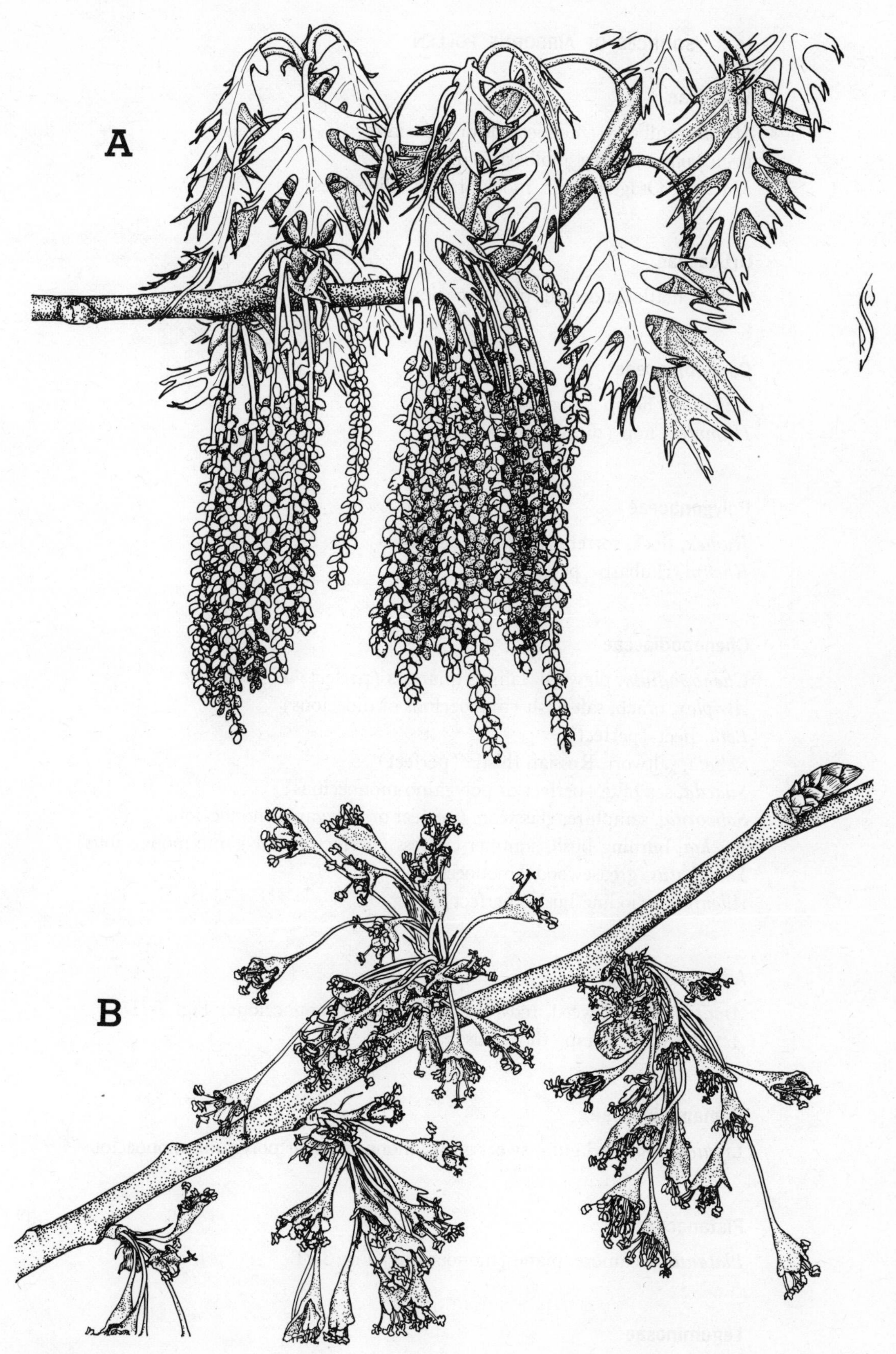

FIGURE 3-3 A, branch of Quercus coccinea with male catkins. B, branch of Ulmus americana with bisexual flowers.

Moraceae

Morus, mulberry (monoecious or dioecious)
Broussonetia, paper mulberry (dioecious)
Maclura, Osage orange, mock orange (dioecious)

Urticaceae

Urtica, nettle (dioecious or monoecious)

Cannabinaceae

Cannabis, hemp (dioecious)
Humulus, hop (dioecious)

Polygonaceae

Rumex, dock, sorrel (perfect or dioecious) Fig. 3-7B
Rheum, rhubarb (perfect)

Chenopodiaceae

Chenopodium, pigweed, lamb's-quarters (perfect) Fig. 3-7C
Atriplex, orach, salt bush (monoecious or dioecious)
Beta, beet (perfect)
Salsola, saltwort, Russian thistle (perfect)
Suaeda, sea blite (perfect or polygamo-monoecious)
Salicornia, samphire, glasswort (perfect or polygamo-monoecious)
Kochia, burning bush, summer cypress (perfect or polygamo-monoecious)
Sarcobatus, greasewood (monoecious)
Allenrolfea, iodine bush (perfect)

Amaranthaceae

Amaranthus, pigweed, red-root (dioecious or monoecious) Fig. 3-7D
Acnida, water hemp (dioecious)

Hamamelidaceae

Liquidambar, red gum, sweet gum (monoecious or polygamo-monoecious)

Platanaceae

Platanus, sycamore, plane (monoecious) Fig. 3-7E

Leguminosae

Acacia, acacia (perfect)
Prosopis, mesquite, screw bean (perfect)

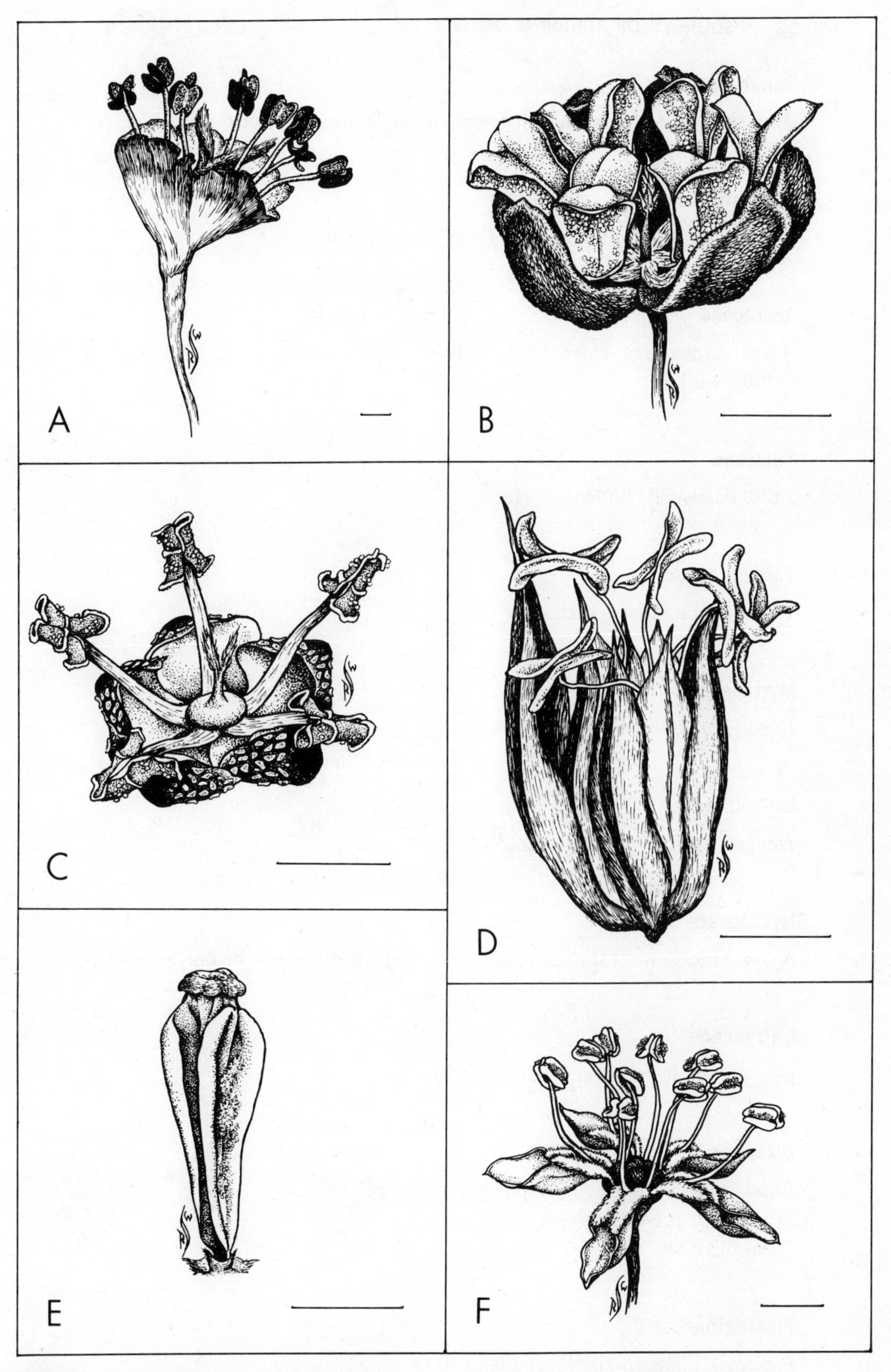

FIGURE 3-4 Flowers and flower parts, all diagramatic. A, bisexual flower of Juncus with six stamens. B, male flower of Populus with many stamens. C, flower of Acer, one unisexual, the other bisexual. D, portion of anther cut to show four sporangia. E, stamen shedding pollen.

Simarubaceae

Ailanthus, tree-of-heaven (dioecious or polygamo-dioecious) Fig. 3-7F

Euphorbiaceae

Ricinus, castor oil plant, castor bean (monoecious)

Aceraceae

Acer, maple, box elder (polygamo-dioecious, rarely perfect) Figs. 3-4C and 3-8A and B

Tiliaceae

Tilia, basswood, linden (perfect)

Tamaricaceae

Tamarix, tamarisk (perfect)

Myrtaceae

Eucalyptus, gum (perfect)

Umbelliferae

Daucus, carrot (perfect) Fig. 3-8C

Nyssaceae

Nyssa, black gum, tupelo, pepperidge (perfect, dioecious, or polygamo-dioecious)

Garryaceae

Garrya, silk-tassel, bear bush (dioecious)

Oleaceae

Fraxinus, ash (dioecious or polygamo-dioecious)
Ligustrum, privet (perfect)
Olea, olive (dioecious or polygamo-dioecious)

Plantaginaceae

Plantago, plantain (perfect) Fig. 3-8D

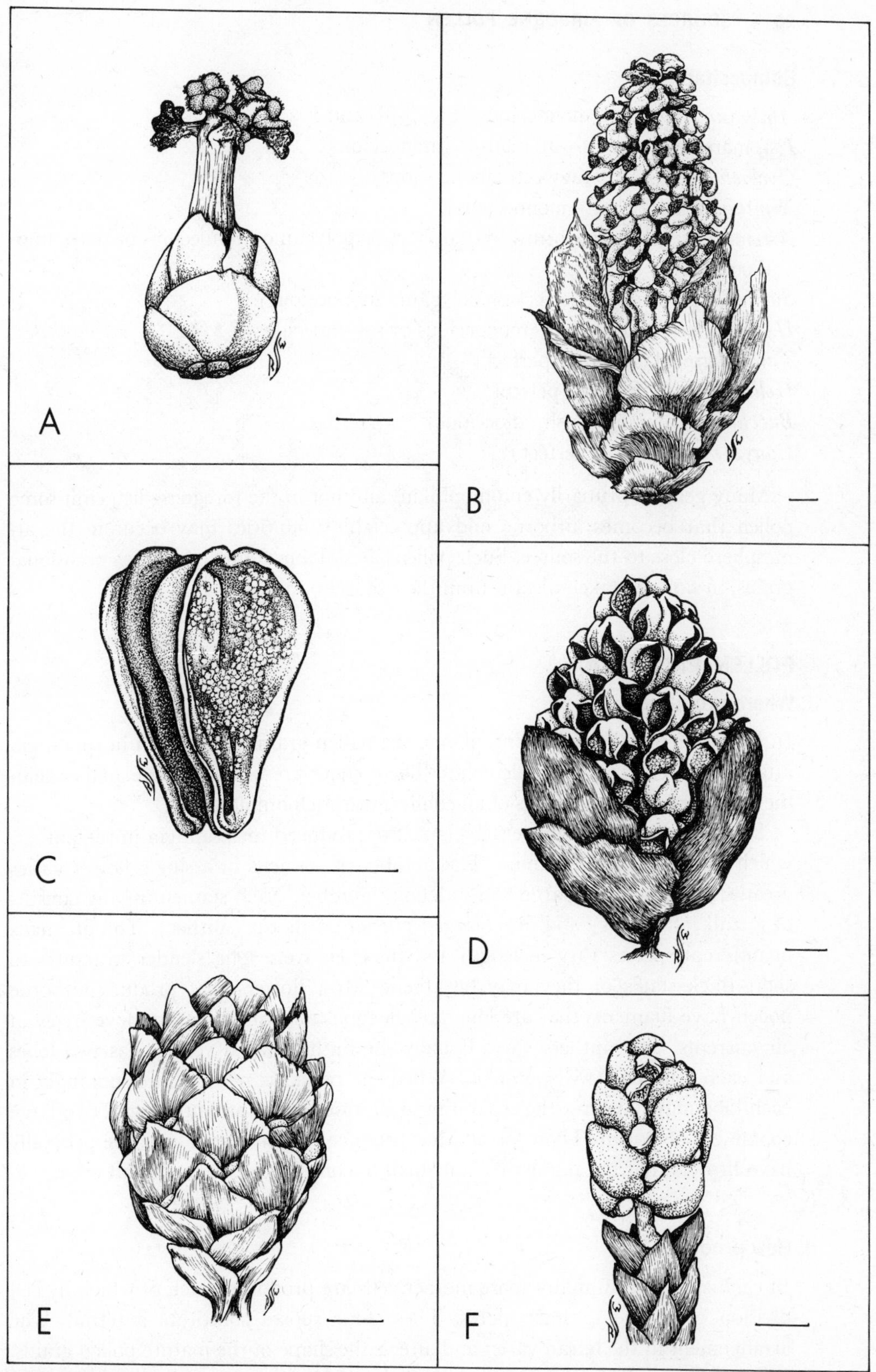

FIGURE 3-5 Male cones of some conifers. A, Taxus canadensis. B, Pseudotsuga menziesii. C, cone scale of Pinus strobus. D, Pinus strobus. E, Sequoia sempervirens. F, Juniperus virginiana. Length of line in millimeters indicates magnification.

Compositae

Ambrosia, ragweed (monoecious) Fig. 3-8E and F
Iva, marsh elder, high-water shrub (monoecious)
Cyclachaena, prairie ragweed (monoecious)
Xanthium, cocklebur (monoecious)
Artemisia, sagebrush, wormwood (perfect or polygamo-monoecious or polygamo-dioecious)
Solidago, goldenrod (perfect or polygamo-monoecious)
Hymenoclea, burro brush (monoecious or sub-dioecious)
Taraxacum, dandelion (perfect)
Helianthus, sunflower (perfect)
Baccharis, groundsel bush (dioecious)
Conyza, horseweed (perfect)

Many genera, primarily entomophilous and not in the foregoing list, emit some pollen that becomes airborne and appreciable quantities may occur in the atmosphere close to the source. Such pollens are seldom found, except as occasional grains, in samples taken distant from their sources.

POLLEN PRODUCTION

Where produced

In the conifers or cone-bearing plants, the pollen grains are formed in sporangia attached to scales of the male cones. These cones are usually much smaller than the female cones. The pollens of all conifers are anemophilous.

The pollen grains of flowering plants are produced in sporangia in the anther, which is a part of the stamen. The number of stamens in a single flower varies greatly; from one to a large and indefinite number. Each stamen usually consists of a stalk (filament) and an enlarged portion at its tip (anther). The filaments of different species vary in length and thickness from long slender structures to short thick stalks or they may be absent. Most flowers that produce airborne pollen have filaments that are long and flexuous allowing them to move freely in air currents. The anthers vary. Usually the anther, when young, has two lobes and each lobe has two sporangia. When the pollen is mature, the sporangia in each lobe may have lost the separating wall, resulting in an anther with two large containers of pollen. Those plants that produce pollen that is airborne generally have flowers that are numerous, but small, inconspicuous, and without odor.

How produced

In each sporangium many spore mother cells are produced, each of which by cell division gives rise to four spores. These four spores constitute a tetrad. The arrangement in the tetrad varies and affects the shape of the mature pollen grains. This is discussed in Chapter 9. At the time of release from the sporangia, the pollen is usually separated into single grains, but in some plants they remain as tetrads (e.g., Ericaceae, *Typha latifolia*). In the sedges (Cyperaceae), the four microspores adhere as a tetrad but three of them abort and shrivel. These

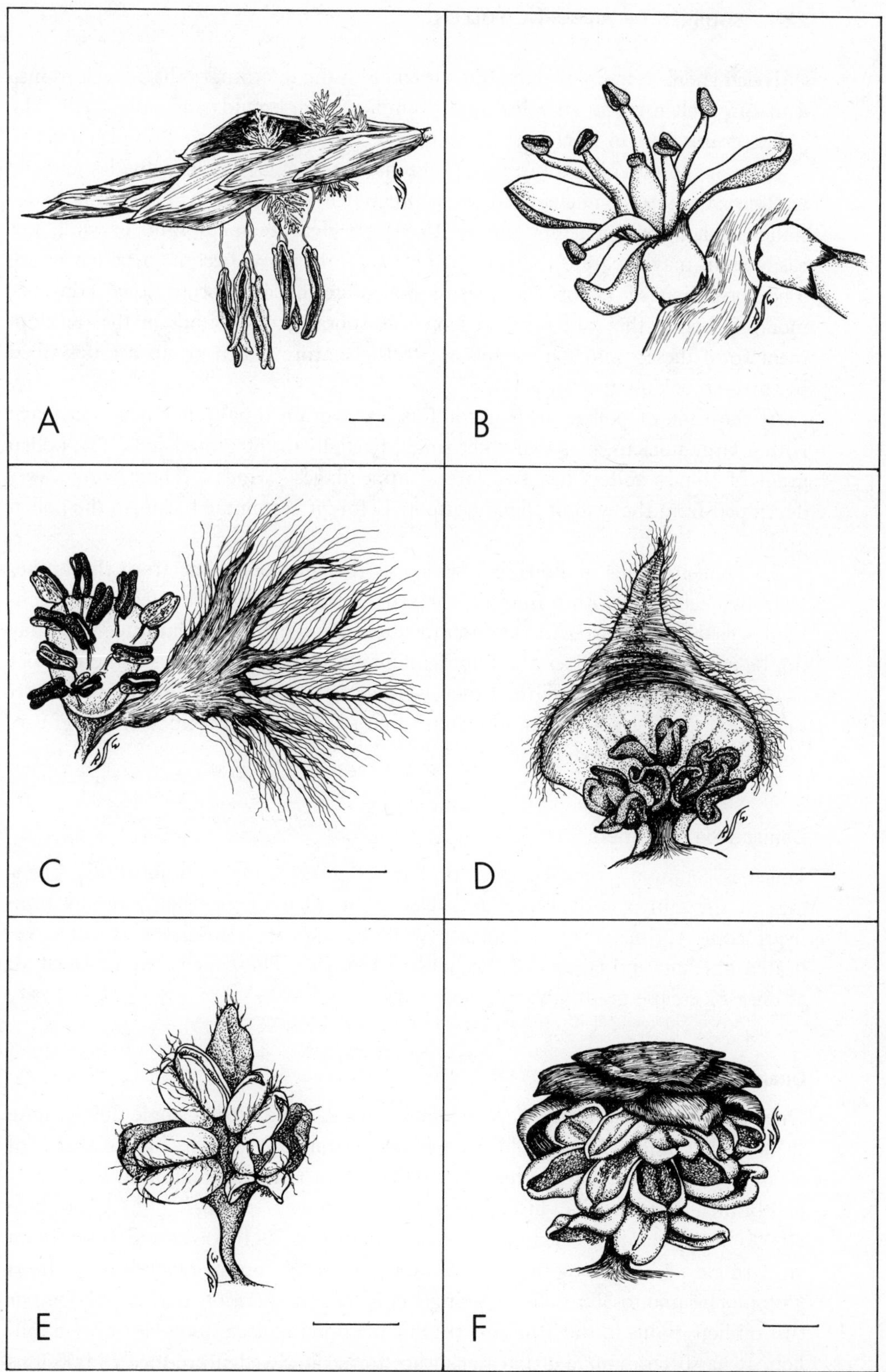

FIGURE 3-6 A, spikelet of flowers of Bromus inermis. B, bisexual flower of Sabal palmetto. C, male flower of Populus tremuloides. D, male flower of Comptonia peregrina. E, male flower of a species of Carya. F, male flower of Betula papyrifera. Length of line in millimeters indicates magnification.

shriveled spores remain as remnants attached to the microspore that develops into a mature pollen grain. In a few plant families (e.g., orchids and milkweeds) the pollen grains remain in clusters (pollinia). These are not carried on air currents.

In each spore, while still in the sporangium, the single nucleus divides and the wall surrounding the nuclei becomes impregnated with a chemically complex substance called sporopollenin (Shaw 1970). A delicate membrane or wall, not easily seen, may separate the nuclei. This spore has now become a pollen grain. The formation of essentially mature pollen grains may occur long (days or months) before they are released from the sporangium. Details of the development from these spores (microspores) to the mature pollen grains are described in any textbook on general botany.

At the time of pollen shed in conifers, each grain usually has two cells, each with a large nucleus, plus two other small, partially disintegrated cells. The pollen grain of some conifers has two lateral appendages (wings). These wings were developed from the wall of the microspore before it germinated to form the pollen grain.

The pollen grains of flowering plants at the time of release from the anther have two cells, one with a nucleus, the other with one or two nuclei. Using techniques usually employed in the identification of pollen, the separating wall may not be evident. The nuclei also may be difficult or impossible to see. The number of nuclei at anthesis (flowering time) is usually constant for the groups of flowering plants. Most trees that produce airborne pollen have two nuclei. The grasses and ragweeds have three nuclei.

Dehiscence of anthers

Pollen is commonly released from the anther through longitudinal slitlike openings in the anther wall. Other methods occur. Dehiscence usually results from hygroscopic shrinkage of the anther wall. Change in humidity may cause repeated opening and closing of the pollen chambers. The pollen may be freed all at once or escape gradually.

Quantity

One stamen of a beech tree (*Fagus*) may have 2,000 grains, a single flower more than 10,000, and a 10-year-old branch system some 30 million. In one anther of a birch tree (*Betula*), there may be 10,000 grains, or 100 million on a 10-year-old branch system. The little sheep sorrel (*Rumex acetosella*) may produce 30,000 in one stamen, which figures to be 180,000 in each flower and 400 million on a single plant. An estimate for a plant of the European black alder (*Alnus glutinosa*) came to 365 billion. A single cone of pine (*Pinus*) may supply one or two million grains to the atmosphere; one of spruce (*Picea*) one-half to two million. It has been estimated that the spruce forests in Sweden produce 75,000 tons of pollen each year (Faegri and Iversen 1964). Each anther of corn (*Zea mays*) contains some 2,000 to 2,500 grains. As there are approximately 7,000 anthers in each tassel, a single plant may shed more than 14 million pollen grains.

A single robust common ragweed plant (*Ambrosia artemisiifolia*) will release

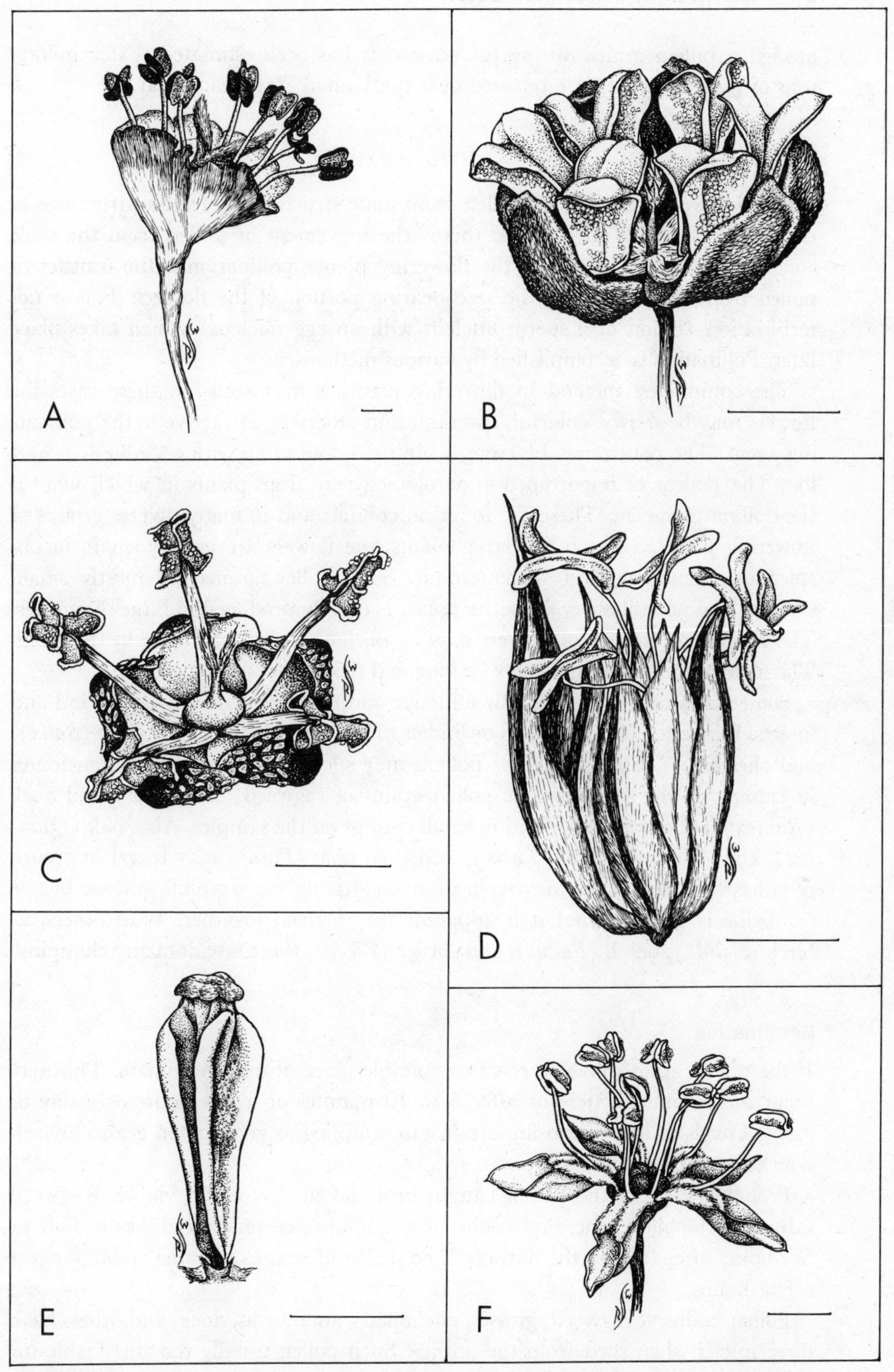

FIGURE 3-7 A, bisexual flower of Ulmus americana. B, bisexual flower of Rumex acetosella. C, bisexual flower of Chenopodium album. D, male flower of Amaranthus retroflexus. E, anther of Platanus occidentalis. F, male flower of Ailanthus altissima. Length of line in millimeters indicates magnification.

around a billion grains during the season. It has been estimated that a million tons of ragweed pollen are released over the United States each year.

POLLINATION

Pollination is the transfer of pollen from male structures to female structures of the same species. With conifers, this is the movement of pollen from the male cones to the female cones. In the flowering plants, pollination is the transfer of pollen from the anthers to the seed-bearing portion of the flowers. This is not fertilization (union of a sperm nucleus with an egg nucleus) which takes place later. Pollination is accomplished by various methods.

The commonest method in flowering plants is by insects. In these cases the flowers may be showy, colorful, fragrant, and otherwise attractive to the pollinating agent. The pollen may be large, sculptured, and often with an adhesive coating. The pollens of importance in aerobiology are from plants in which wind is the pollinating agent. These are found in conifers and in many diverse groups of flowering plants. In such flowering plants, the flowers are usually small, inconspicuous, numerous, and without odor. Such pollen grains are mostly small, smooth, and nonadhesive. Airborne pollen is usually produced in large quantities. The flowers or clusters of flowers may be on long stalks that wave in the wind. The filaments of the anthers may be long and flexuous.

Some airborne pollen is slightly adhesive and may be carried by both wind and insects. Examples are: basswood or linden (*Tilia*), maple (*Acer*), willow (*Salix*), and chestnut (*Castanea*). These pollens may stick together and are often found in clumps on the samples. The pollen grains of ragweed (*Ambrosia*) and dock (*Rumex*) are sometimes found in small clumps on the samples. Also, oak (*Quercus*), elm (*Ulmus*), larch (*Larix*), and even pine (*Pinus*) may travel in groups of adhesive grains. They may separate upon striking the sampling surface but be so obviously grouped that it is apparent they arrived together. With others, as birch (*Betula*), beech (*Fagus*), and spruce (*Picea*), there is seldom any clumping.

Germination

If the pollen grain comes to rest in a suitable place, it may germinate. This may occur almost immediately or after 5 to 10 minutes or a few hours or a day or more. This is of little or no importance in sampling as germinated grains are seldom encountered in the samples.

Probably most of the pollen caught in aerial surveys is not viable. Ragweed (*Ambrosia*) pollen generally reaches it maximum germination ability in four or five hours after leaving the anthers. The pollen of grasses remains viable for just a few hours.

Pollen grains of ragweed, grasses, chenopods, amaranths, dock, and others have three nuclei when shed from the anther. Such pollen usually remains viable for but a short time. Many kinds of pollen have only two nuclei when shed. They usually are viable for longer periods: several days or more than a year. These include cattail (*Typha*), oak (*Quercus*), beech (*Fagus*), birch (*Betula*), poplar (*Populus*), maple (*Acer*), and palms (*Cocos, Phoenix,* and many others).

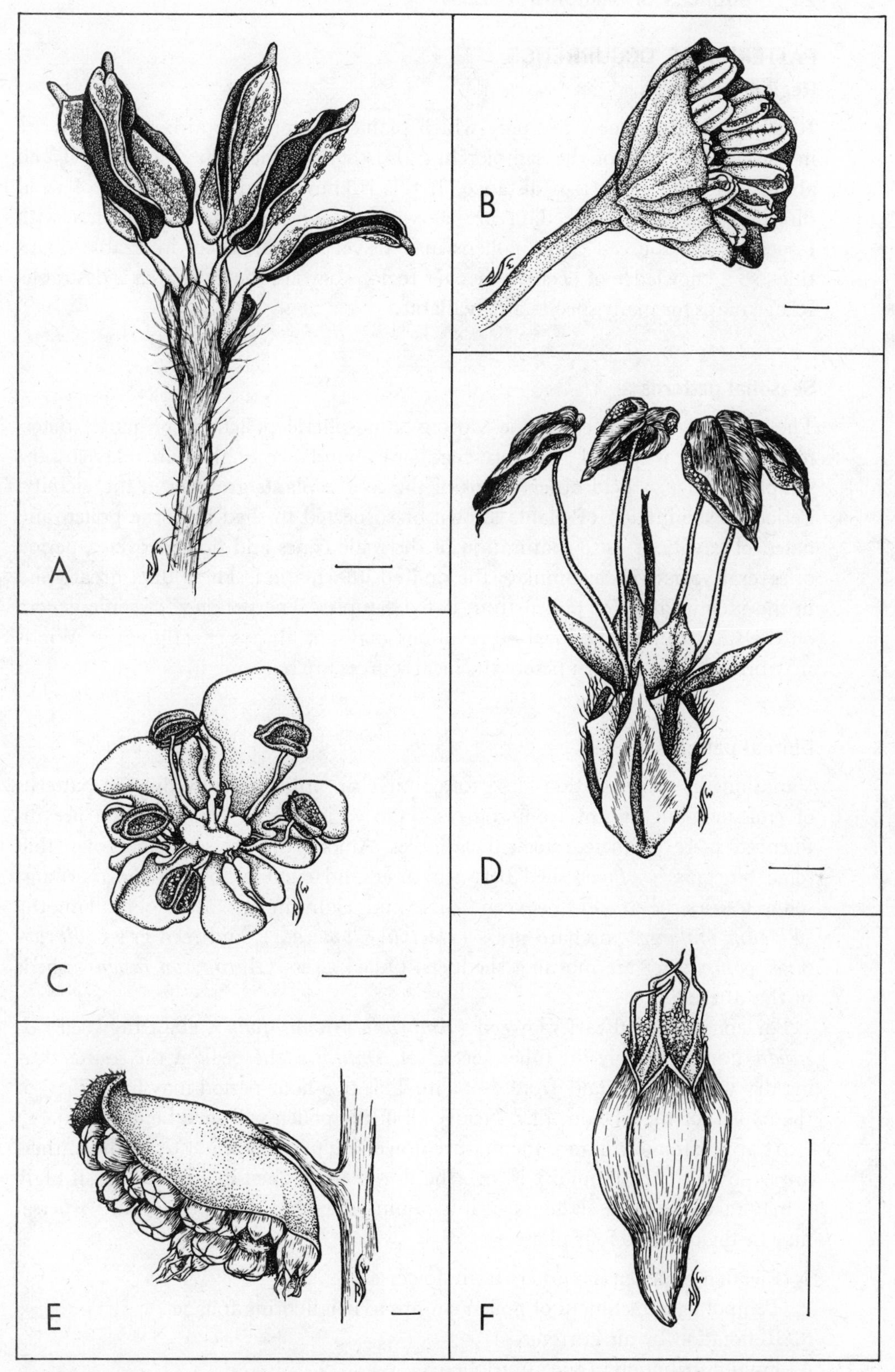

FIGURE 3-8 A, male flower of Acer negundo. B, young male flower of Acer saccharum. C, bisexual flower of Daucus carota. D, bisexual flower of Plantago lanceolata. E, male flower cluster of Ambrosia artemisiifolia. F, male flower of Ambrosia artemisiifolia. Length of line in millimeters indicates magnification.

PATTERNS OF OCCURRENCE

Regional patterns

It is important to know not only which plants that produce airborne pollen are in the general area of the sampler but also their positions in relation to it and abundance with respect to distance. If this is known and daily records of wind direction are kept, these data are of value in understanding the catch with respect to the sources. Some pollens may travel long distances in sizable quantities, so a knowledge of occurrence over regions beyond the local area is desirable. Range maps for many species are available.

Seasonal patterns

The seasonal occurrence of the various atmospheric pollens is primarily determined in two ways: (1) the presence and abundance of these as taken by the samplers and (2) field observations of the source plants growing in the vicinity. Periodic examination of plants known or suspected to shed airborne pollen and dated observations as to maturation of the male cones and flowers over a period of several years will accumulate the desired information. These data greatly aid in the identification of pollen from dated samples. The presence of some genera on the samples and the local observations may not always exactly agree. Winds may bring in some pollen before the local sources are releasing it.

Diurnal patterns

Variations in concentration close to a source are influenced by diurnal patterns of emission, but distant from sources these variations in concentration are influenced more by meteorological variables. Among the grasses it is known that some bluegrasses (*Poa*) shed between three and eight o'clock in the morning; some fescues (*Festuca*) between three and eight in the afternoon. Timothy (*Phleum pratense*), orchard grass (*Dactylis glomerata*), and reed grass (*Phragmites communis*) are morning shedders. Quack grass (*Agropyron repens*) sheds in the afternoon.

For common or dwarf ragweed (*Ambrosia artemisiifolia*), giant ragweed (*A. trifida*), and probably the other species of *Ambrosia,* the peak at the source during dry weather is usually from 7–9 a.m. This two-hour period may have 75% of the pollen for the 24-hour day. Nearly all of the pollen will shed before noon.

At approximately 6 a.m. the mature flowers change shape; at 6:30 the anthers are exposed. If the humidity is low, the flowers may open in 15 minutes; if high, it may take two or three hours; if it is raining they may not open at all. Dispersal may be divided into four phases:

1. Ejection of pollen in clusters from flowers.
2. Temporary attachment of pollen clumps to neighboring foliage.
3. Reflotation by air currents.
4. Final distribution in the atmosphere.

Little is known about the diurnal patterns of emission of tree pollens. Pollen shed from the catkins of the white birch (*Betula papyrifera*) apparently is completed within a few days for each tree, without a marked diurnal pattern. Jack pine (*Pinus banksiana*) pollen diurnal patterns appear to be correlated with humidity. The apple is primarily entomophilous, but appreciable quantities of pollen may be in the air close to the trees; the highest concentrations are in the afternoon.

References

Durham, O.C., and J. Borenstine. 1963. Pollen allergens. Pages 41–74 *in* F. Speer, ed. The allergic child. Hoeber Med. Div., Harper and Row Publ., New York.

Eames, A.J. 1961. Morphology of the angiosperms. McGraw-Hill Book Co., New York. 518 p.

Faegri, K., and J. Iversen. 1964. Textbook of pollen analysis. 2nd rev. ed. Hafner Publ. Co., New York. 237 p.

Gottlieb, P.M., and E. Urbach. 1943. The distribution and pollination times of the important hay fever-producing plants in the United States. J. Lab. Clin. Med. 28:1053–1070.

Hardin, J.W. 1968. Commercial herbs, roots and pollens of North Carolina. N.C. Agr. Exp. Sta. Bull. 435. 76 p.

Meeuse, B.J.D. 1961. The story of pollination. Ronald Press, New York. 243 p.

Percival, M.S. 1965. Floral biology. Pergamon Press, New York. 243 p.

Shaw, G. 1970. Sporopollenin. Pages 31–58 *in* J.B. Harborne, ed. Phytochemical phylogeny. Academic Press, New York.

Solomon, W.R., O.C. Durham, and F.L. McKay. 1967. Pollens and the plants that produce them. Pages 340–397 *in* J.M. Sheldon, R.G. Lovell, and K.P. Mathews, A manual of clinical allergy. 2nd ed. W.B. Saunders Co., Philadelphia.

Wodehouse, R.P. 1935. Pollen grains. McGraw-Hill, New York. 574 p. (Reprinted in 1959 by Hafner Publ. Co., New York).

Wodehouse, R.P. 1971. Hayfever plants. 2nd rev. ed. Hafner Publ. Co., New York. 280 p.

4

REFERENCE COLLECTIONS

A collection of pollen grains for comparison with unknown grains is a necessity. These may be preserved and filed in several ways. Each species should be represented by collections from different areas, different habitats, and different years. They are easily collected from local plants. Pollens of the common species represented in the atmosphere are available from several commercial sources. To broaden the reference collection, dried flowers and staminate cones or prepared slides may be traded among palynologists (those who study pollens and other spores). For some of these samples of pollen, it may be desirable to collect, preserve, and file enough of the source plant to allow critical identification. Such a voucher is useful when the identity of a pollen sample is questioned.

BULK STORAGE

A supply of various kinds of pollens may be stored to facilitate making check-slides when needed or for exchange. These may be preserved dry in vials or packets or in suitable liquid media. Glycerin jelly or glycerin are commonly used. Glacial acetic acid, commonly employed by paleopalynologists, is not suitable unless the pollen is to be acetolyzed for special studies.

Packets

A few dried flowers containing mature pollen may be enclosed in folded glassine paper (powder paper) and placed into a coin envelope (Fig. 4-1). The name of the plant (genus and species) is placed on the outside of the envelope, allowing the collections to be filed alphabetically. Other data, such as date, locality, and reference to a voucher collection may be included. A drying agent should be added. A suitable method is as follows:

1. Fold a sheet of glassine paper (approximately 4.5 × 5.5 inches) through the middle to make a trough 4.5 inches long.
2. Shake fresh pollen, when abundant, into the trough or place clipped anthers or whole flowers (even small branches) between the sides of the folded paper.
3. Place crystals of a drying agent, such as silica gel desiccant, into an embedding bag. These bags are commercially available for use in histological techniques.

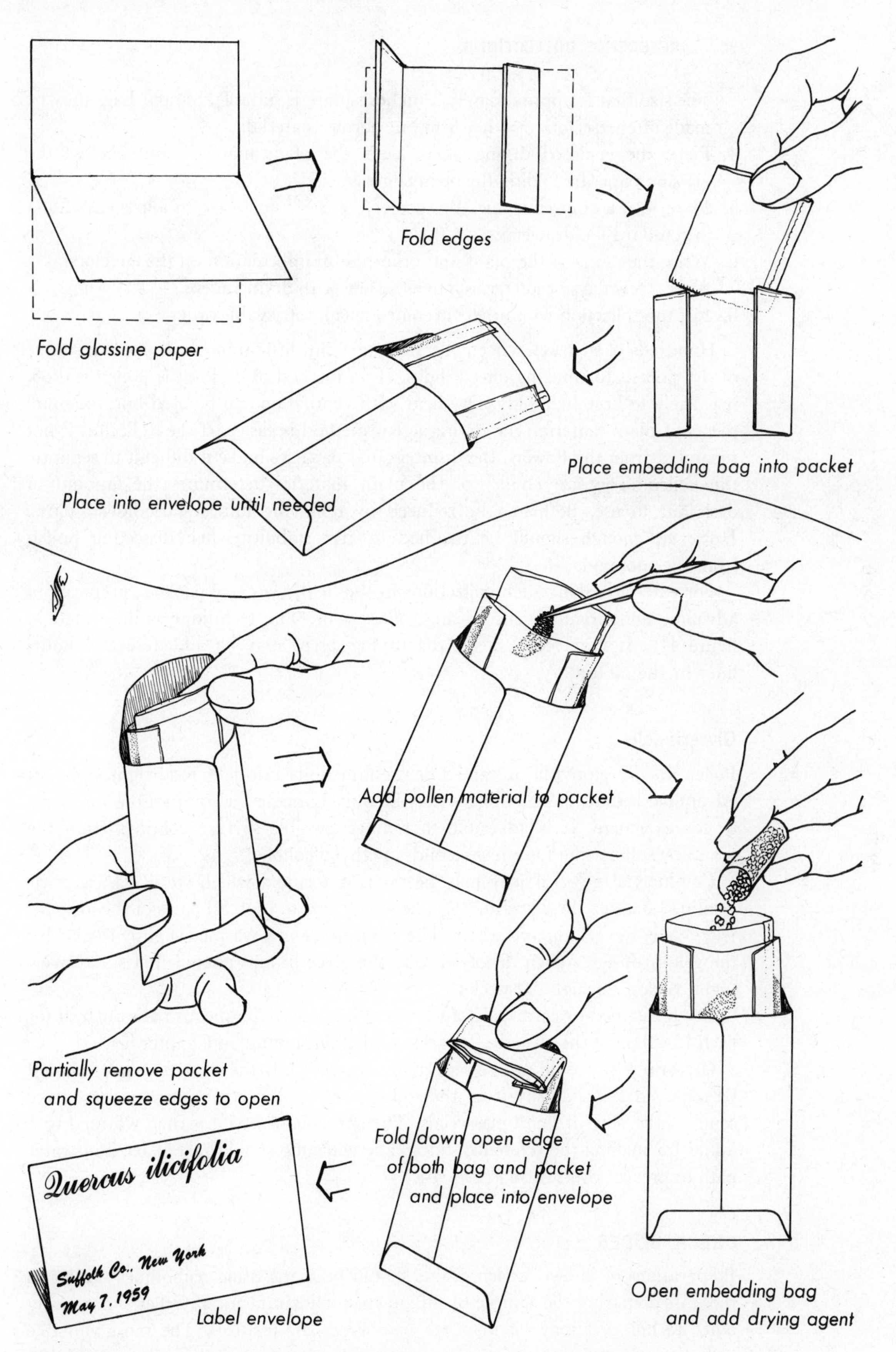

FIGURE 4-1 Preparation and use of envelopes for pollen collections.

The size that is approximately 2 inches square is suitable. Similar bags may be made of cheesecloth or other firm but porous material.

4. Place the enclosed drying agent with the plant material and enclose the glassine paper by folding the open edges.
5. Place into a coin envelope. We prefer the 3 × 5-inch size to allow easy filing in standard file drawers.
6. Write the name of the plant and other useful information on the envelope.
7. After a few days or longer, remove the bag with drying agent.
8. File the collection with others in some logical, retrievable order.

Hundreds of such reference collections take but little room and supply a source of dry pollens for making check slides. If no method of drying the pollen is used, it is likely to become moldy. Dry heat with ventilation can be used but, for small pieces of plant material, the drying agent method is easier. If the desiccant is not separated from the flowers, the disintegrated particles make it difficult to separate the pollen. The size (bulk) of the plant material determines the amount of desiccant to use. Both can be reduced by removing unnecessary flower parts. However, enough should be retained to give stability when dissecting pollen from the sporangia.

For ease of making such collections in the field, we carry packets, prepared in advance, and drying agent in small glass vials. The technique is illustrated in figure 4-1. If more convenient, the drying agent may be added several hours later in the laboratory.

Glycerin jelly

Pollen may be stored in unstained or prestained glycerin jelly in airtight vials. An advantage is that reference slides may be quickly made for comparison, for filing, or for exchange. It is advisable to remove possible surface coatings from the grains by submersion for a few seconds in ethyl alcohol.

Commercial glycerin jelly may be used. A formula which we like is: 2 parts gelatin, 12 water, 11 glycerin, 2% phenol (1 part to each 50 parts of the glycerin jelly). Mix the gelatin and water. The mixture may be warmed slightly to dissolve the gelatin faster. When dissolved, add the glycerin and phenol. Let stand overnight. Strain through cheesecloth.

For prestained glycerin jelly, add a drop of saturated aqueous basic fuchsin for each 15-20 ml of the glycerin jelly. See section on staining in Chapter 8.

Glycerin jelly should not be repeatedly heated. Only the amount needed should be removed as solid jelly from the storage bottle. A useful technique is to store some of the jelly in small glass vials. The contents of a vial is then warmed to a liquid for making the reference slides. The warming should take place in a water bath to prevent overheating (Fig. 4-2).

CHECK SLIDES

Preparations of known pollen grains should be made using techniques similar to those in preparing the sample of unknowns for identification. Other slides, using other techniques, may be made to clarify certain features. The most valuable collections include a variety of preparations. However, the most useful are generally those that simulate the technique used for processing the samples of

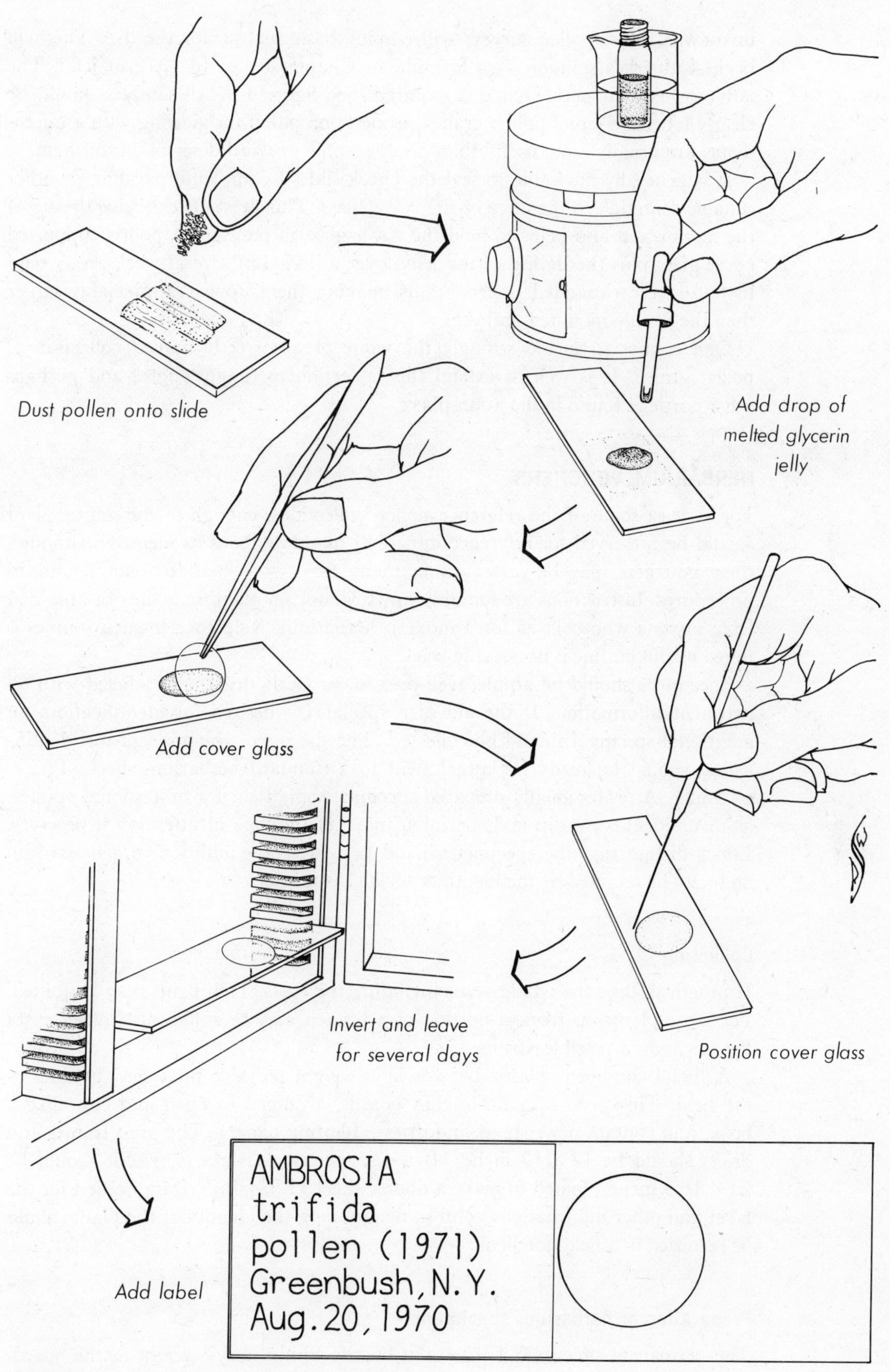

FIGURE 4-2 Preparation of check slides.

unknowns. Most pollen surveys will employ basic fuchsin for the dye. This will be in Calberla's solution (see formula in Chapter 8) or in glycerin jelly. The latter is better if permanence is required. See figure 4-2. All samples should be clearly labeled. Actual pollen grains, properly prepared for viewing with a microscope, are usually more useful than photographic or word descriptions of them.

It is generally advisable to seal the check slides by applying paraffin, or other suitable material, to the edge of the cover glass. This prevents excessive drying of the mount and also helps to hold the cover glass in position. A poorly supported cover glass plus the drying of the jelly, over a period of several weeks, may tend to flatten the expanded pollen grains, making them appear appreciably larger than the freshly mounted grains.

One cannot stress too strongly the value of a correctly named collection of pollen grains. It is well to extend this collection to fungus spores and perhaps other particles found in the atmosphere.

HERBARIUM VOUCHERS

For at least some of the reference pollen collections, enough of the source plant should be preserved for reference in case of question about its identity. Although these vouchers may be preserved in many ways, it is best to follow standard procedures. Instructions are found in any textbook on plant taxonomy or obtained from anyone who collects for a modern herbarium. Reference to such sources is urged as this outline is necessarily brief.

Specimens should be ample, well-pressed, properly dried, and labeled with all pertinent information. If the aid of a specialist is desired for identification, an additional specimen should be collected. The specimen which is sent for identification should be ready for attachment to a standard herbarium sheet (11.5 × 16.5 in.). A professionally prepared specimen should elicit a professional opinion as to its identity; a slipshod specimen may get the type of attention it deserves. Ethics dictate that the specimen should be a welcome addition to a herbarium and may be retained by the identifier for his institution.

Collecting

If practical, take the whole plant including the roots. Tall plants may be folded. For large plants, as trees, a portion of a branch may be sufficient. Collect extra flowers and (if possible) fruits.

A metal container, plastic bag, or light-weight portable press may be used in the field. This press may be of thin boards, arranged to open and close like a book, and contain newspapers and driers (blotting paper). The press frames and driers should be 12 × 18 inches. If newspaper stock is cut to size, it should be 23 × 16.5 inches, folded to make a double sheet 11.5 × 16.5. Data needed for the label and other information useful in determining the identity of the plant should be recorded in a field notebook.

Preparation of herbarium specimens

The permanent press may be two flat boards with a heavy weight or the boards (or frames of hardwood slats) pressed together with straps, rope, or chain. The

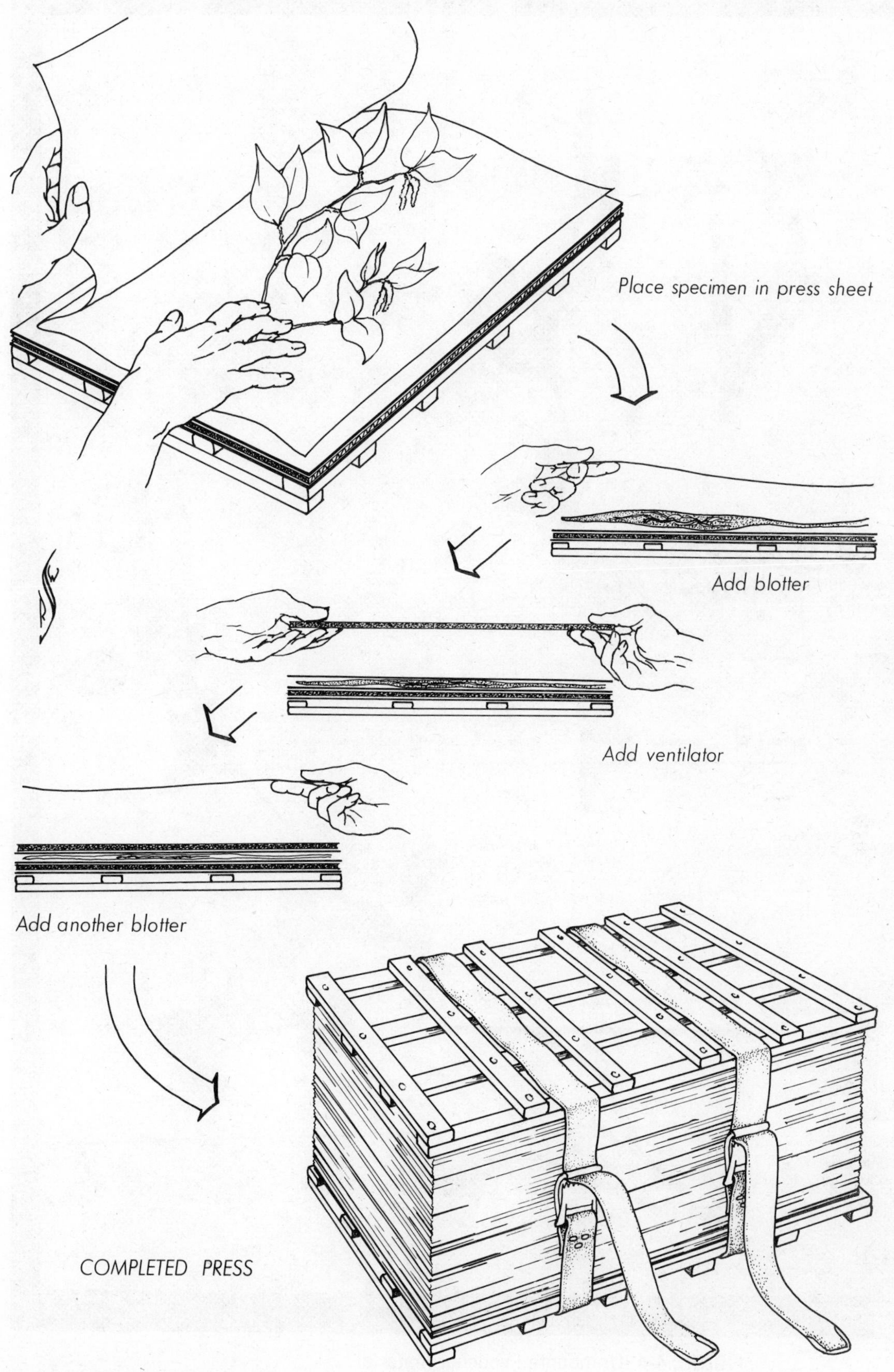

FIGURE 4-3 Preparation of herbarium vouchers.

FIGURE 4-4 Unmounted voucher material.

FIGURE 4-5 Collection ready for mounting or for exchange.

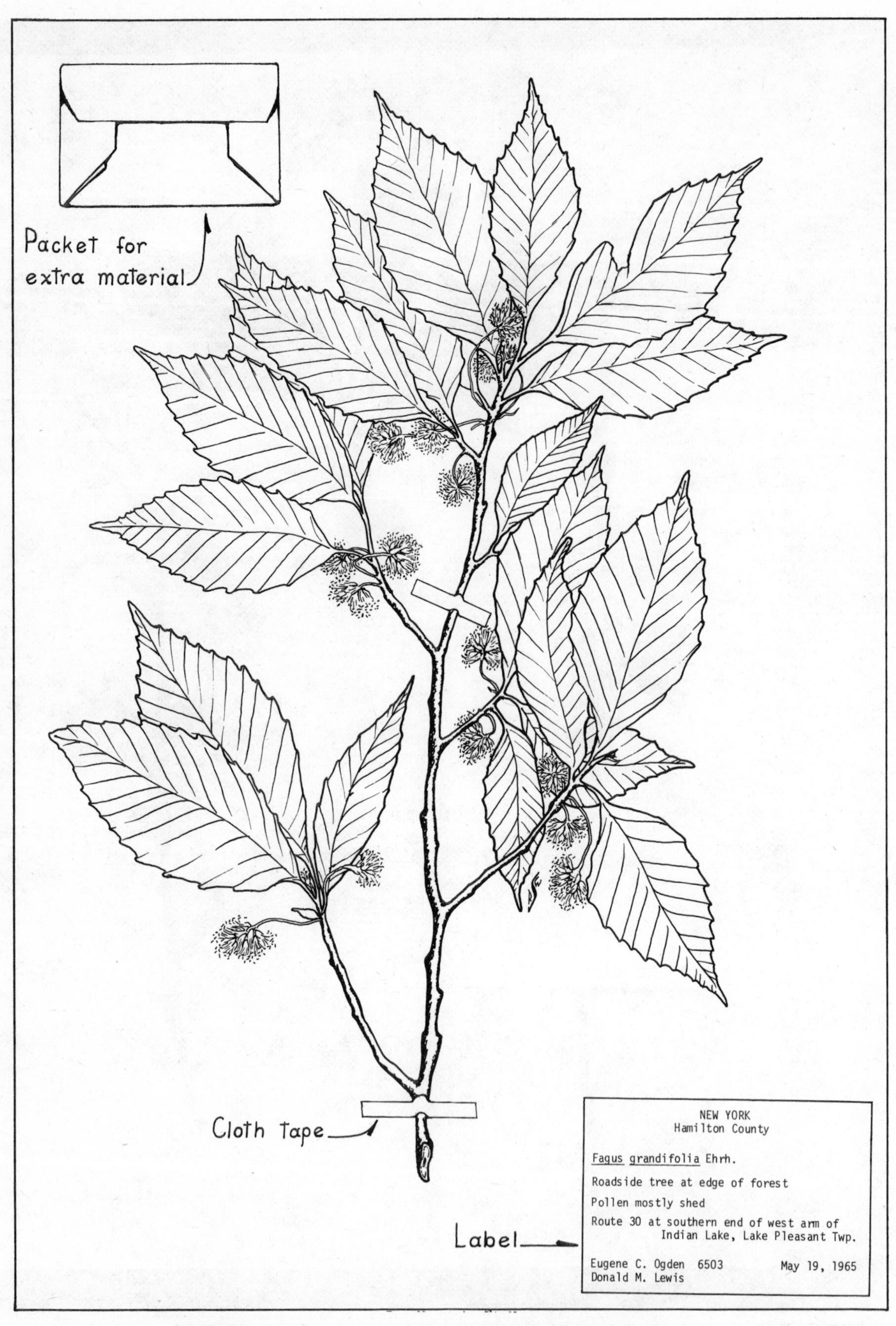

FIGURE 4-6 Parts of a typical mounted voucher collection.

FIGURE 4-7 Mounted herbarium sheet.

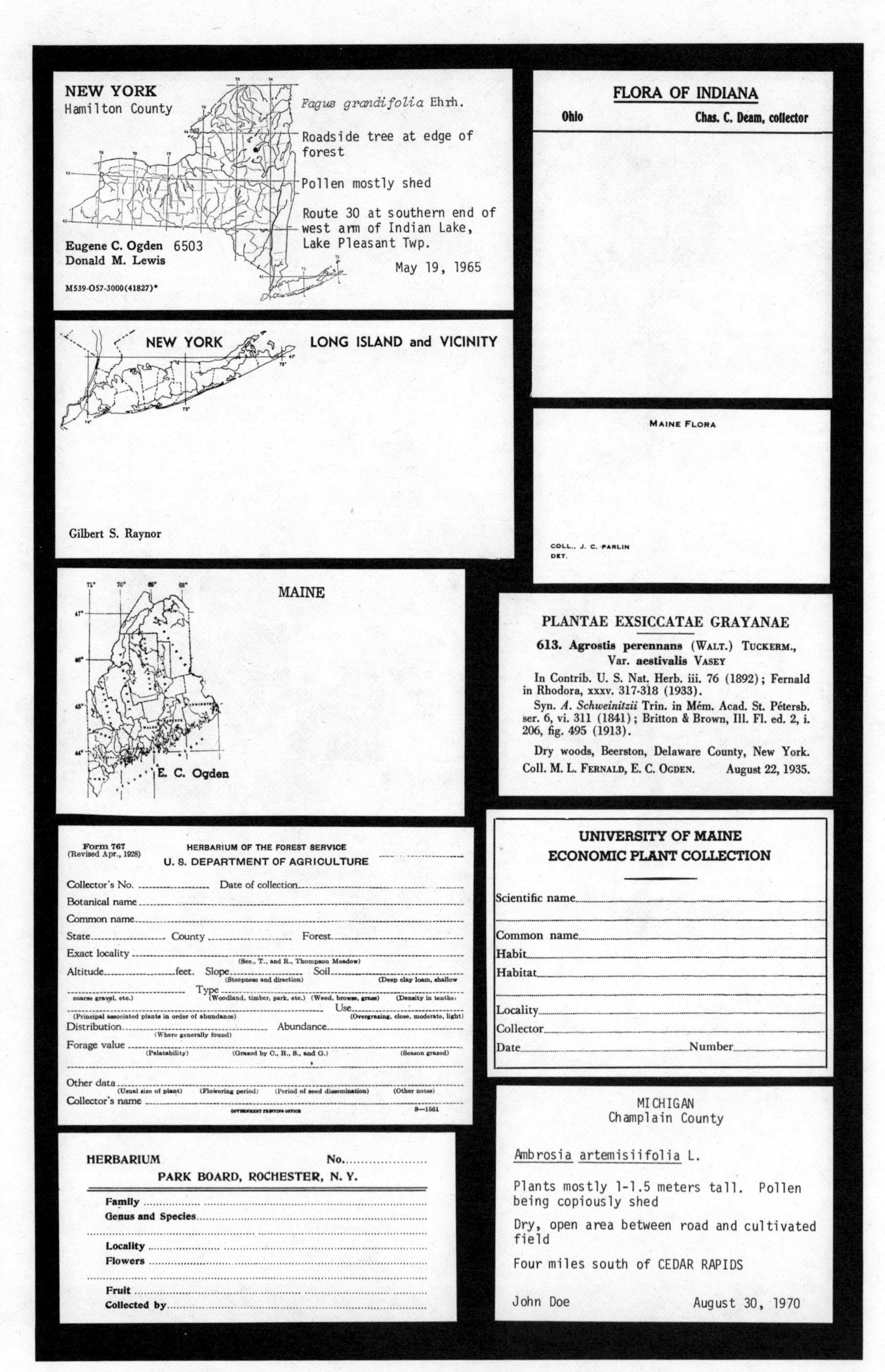
NEW YORK
Hamilton County

Fagus grandifolia Ehrh.

Roadside tree at edge of forest

Pollen mostly shed

Route 30 at southern end of west arm of Indian Lake, Lake Pleasant Twp.

Eugene C. Ogden 6503
Donald M. Lewis

May 19, 1965

M539-O57-3000(41827)*

FLORA OF INDIANA

Ohio — Chas. C. Deam, collector

NEW YORK — LONG ISLAND and VICINITY

Gilbert S. Raynor

MAINE FLORA

COLL., J. C. PARLIN
DET.

MAINE

E. C. Ogden

PLANTAE EXSICCATAE GRAYANAE

613. Agrostis perennans (WALT.) TUCKERM., Var. **aestivalis** VASEY

In Contrib. U. S. Nat. Herb. iii. 76 (1892); Fernald in Rhodora, xxxv. 317-318 (1933).

Syn. *A. Schweinitzii* Trin. in Mém. Acad. St. Pétersb. ser. 6, vi. 311 (1841); Britton & Brown, Ill. Fl. ed. 2, i. 206, fig. 495 (1913).

Dry woods, Beerston, Delaware County, New York.
Coll. M. L. FERNALD, E. C. OGDEN. August 22, 1935.

Form 767
(Revised Apr., 1928)

HERBARIUM OF THE FOREST SERVICE
U. S. DEPARTMENT OF AGRICULTURE

Collector's No. .... Date of collection ....
Botanical name ....
Common name ....
State .... County .... Forest ....
Exact locality .... (Sec., T., and R., Thompson Meadow)
Altitude .... feet. Slope .... (Steepness and direction) Soil .... (Deep clay loam, shallow coarse gravel, etc.)
.... Type .... (Woodland, timber, park, etc.) (Weed, browse, grass) (Density in tenths)
.... (Principal associated plants in order of abundance) Use .... (Overgrazing, close, moderate, light)
Distribution .... (Where generally found) Abundance ....
Forage value .... (Palatability) (Grazed by C., H., S., and G.) (Season grazed)
Other data .... (Usual size of plant) (Flowering period) (Period of seed dissemination) (Other notes)
Collector's name ....

GOVERNMENT PRINTING OFFICE 8—1561

UNIVERSITY OF MAINE
ECONOMIC PLANT COLLECTION

Scientific name ....
Common name ....
Habit ....
Habitat ....
Locality ....
Collector ....
Date .... Number ....

HERBARIUM No. ....
PARK BOARD, ROCHESTER, N. Y.

Family ....
Genus and Species ....
Locality ....
Flowers ....
Fruit ....
Collected by ....

MICHIGAN
Champlain County

Ambrosia artemisiifolia L.

Plants mostly 1-1.5 meters tall. Pollen being copiously shed

Dry, open area between road and cultivated field

Four miles south of CEDAR RAPIDS

John Doe — August 30, 1970

FIGURE 4-8 Samples of labels for herbarium collections.

fillers for the press consist of the press sheets (newspaper stock), driers (blotters), and ventilators (corrugated cardboard). Placing the specimen in the press must be done with care as appearance of the final specimen depends on how it is pressed and dried. Special techniques are needed for some plants that may be fleshy, thorny, sticky, juicy, etc. The order of placement in the press is: ventilator, drier, specimen in press sheet, drier, ventilator, etc. (Fig. 4-3). Sufficient pressure is required to keep the specimens flat until dry. Drying may be by natural (sun) or artificial heat.

Herbarium

Every state in the United States has at least one modern herbarium. Nearly 100 herbaria in North America have collections of 50,000 or more specimens. Some institutions have collections of more than a million. Contacts with staff members at a small or large herbarium, botanic garden, or college botany department should be rewarding. A small private herbarium is inexpensive and a useful tool in pollen survey work. The specimens may be filed unmounted in the press sheets (Figs. 4-4 and 4-5) or mounted on standard herbarium sheets (Figs. 4-6 and 4-7). As much care should be given to preparing a proper label, containing essential and useful information, as to preparing the plant specimen (Fig. 4-8). Specimens for identification by specialists are usually supplied unmounted, accompanied by the labels.

References

Benson, L. 1957. Plant classification. D.C. Heath & Co., Boston. *See* Chapter 12, Preparation and preservation of plant specimens.

Core, E.L. 1955. Plant taxonomy. Prentice-Hall, Inc., Englewood Cliffs, N.J. *See* Chapter 8, Gardens and herbaria.

Davis, P.H., and V.H. Heywood. 1963. Principles of angiosperm taxonomy. D. Van Nostrand, Princeton, N.J. *See* Chapter 8, Field, herbarium and library.

Kelly, J.W. 1928. Methods of collecting and preserving pollen for use in the treatment of hay fever. U.S. Dep. Agr. Circ. 46. 9 p.

Ketchledge, E.H. 1970. Plant collecting: a guide to the preparation of a plant collection. N.Y. State College Forestry, Syracuse. 21 p.

Lawrence, G.H.M. 1951. Taxonomy of vascular plants. Macmillan Co., New York. *See* Chapter 11, Field and herbarium techniques.

Paetzel, M.M. 1967. Hints for pollen collectors. Greer Drug and Chem. Corp., Lenoir, N.C. 36 p.

Porter, C.L. 1967. Taxonomy of flowering plants. 2nd ed. W.H. Freeman & Co., San Francisco. *See* Chapter 4, Field and herbarium methods.

Savile, D.B.O. 1962. Collection and care of botanical specimens. Can. Dep. Agr. Res. Br., Publ. 1113. 124 p.

Smith C.E., Jr. 1971. Preparing herbarium specimens of vascular plants. U.S. Dep. Agr. Res. Serv. Inf. Bull. 348. 29 p.

Veldee, M.V., ed. 1948. Specifications recommended as guides in the collection and preservation of pollens. Ann Allergy 6:56–60. Also published in J. Allergy 19:210–214.

5
SAMPLING PRINCIPLES

Methods used in sampling the atmosphere for airborne pollens are identical in principle and often similar in practice to those used for sampling many other small solid and liquid particles such as air pollutants, meteorological tracers, condensation nuclei, and even cloud and fog droplets. Although a great number of sampling devices are in use, all operate on only a few basic principles which are described and illustrated below. Each principle has both advantages and disadvantages, and since some are more suitable than others for collecting particles in the pollen size range, each will be evaluated for this purpose. The four basic sampling methods to be discussed are gravitational settling, impaction, suction, and grab sampling. Instruments using these methods are described in Chapter 6.

GRAVITATIONAL SETTLING

Exposure of a horizontal surface on which particles can settle by gravity is the simplest method of collecting airborne pollen and the method still most frequently used. In theory, particles simply settle at their terminal velocity and are retained by an adhesive on the sampling surface. The terminal velocity of a small, smooth, spherical particle can be computed from Stoke's equation which is given in many physics texts. The equation states that terminal velocity is proportional to the square of the particle radius, the particle density, and the acceleration due to gravity and inversely proportional to the viscosity of the air or other medium through which the particle falls. The equation is not exact for rough or nonspherical particles but should not be greatly in error for most pollens.

In completely calm or very stable air (air lacking turbulence), this concept of gravitational settling is valid since gravity is the predominant depositing mechanism. However, even in calm air, the volume of air sampled is proportional to the settling rate and may differ for each type of particle collected. For example, if two types of pollen grains are collected simultaneously, one of which has twice the settling rate of the other, the heavier particles will have fallen out from twice the height of the other and twice as many will be taken if both were originally present in the same concentration and distributed vertically in the same manner (Fig. 5-1A). Unless the respective settling rates are known and the counts adjusted, a false impression of their relative abundance will be obtained.

If the air is not calm but remains nonturbulent, particles do not settle vertically but descend at an angle determined by their terminal velocity and the wind speed (Fig. 5-1B). The volume sampled now changes with wind speed since more air flows across the sampler as the speed increases. In addition, the upwind edge of the sampler may cause a divergence of the air flow which will cause some of the approaching particles to be carried over the sampling surface. This aerodynamic rise will vary with wind speed and, if the sampler is not shaped symmetrically in the horizontal, the error caused by it will also vary with the direction of air flow across the sampler. These effects have been verified experimentally (Ogden and Raynor 1960).

In turbulent air, conditions become still more complicated. Capture takes place not only by settling but also by turbulent impingement in which downward components of air motion impel particles toward the sampling surface (Fig. 5-1C). These particles have enough momentum to penetrate the thin layer of still air above the sampling surface and are captured by the adhesive coating. Durham (1944) demonstrated the importance of impingement, compared to settling, by collecting about half as much pollen on the bottom of a slide as on the top. Under such circumstances, other mechanisms such as electrostatic and thermal forces may assist in particle capture, particularly for very small particles, but these forces are not normally important for particles as large as pollen.

Thus, the collection efficiency of a "gravity" sampler can be a complex function of particle size, wind speed, wind direction, and turbulence, as well as particle concentration. It is, therefore, impossible to define the volume of air sampled or to compute the concentration of particles in that air. Moreover, counts are not comparable from one time or place to another unless meteorological conditions are identical. At best, such samplers give an indication of the types of particles present and a very rough measure of their abundance.

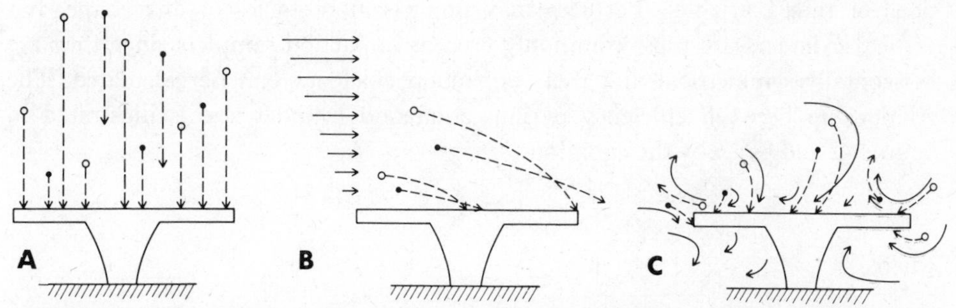

FIGURE 5-1 Diagrams illustrating mechanisms of particle collection by "gravity slide" sampler. Solid arrows represent air trajectories and dashed arrows particle trajectories. A, gravitational settling or sedimentation in calm air. Larger particles fall more rapidly than small particles which may not descend to sampling surface during a finite sampling period. B, settling in a wind speed increasing with height. Trajectories become less horizontal and more vertical as particle descends to layers of decreasing wind speed. Larger particles still settle more rapidly than small particles. C, settling in turbulent air. Particles are collected by turbulent impingement. Particles generally follow eddy motions, but their paths are modified slightly by gravity.

If a horizontal sampling surface is exposed on the ground, it does give a measure of deposition per unit area on that particular surface. However, this gives little information on the concentration in the air above. Deposition on a dissimilar surface nearby may be much different.

IMPACTION

Since wind speeds are generally much greater than gravitational settling rates, most small airborne particles travel a nearly horizontal course. Their mass and velocity give them an inertial force which resists changes in speed and direction. When a particle approaches a physical obstacle, the air molecules surrounding the particle divert and flow around the obstacle. If the particle has sufficient inertia, it will continue on its original course or on a path somewhere between this and the path of the air molecules and may strike the obstacle. In the atmosphere, the efficiency of impaction (the percentage of particles approaching an obstacle that actually strike it) is a direct function of the size, mass, and velocity of the particle and an inverse function of the size of the obstacle. Equations describing the impaction process are mathematically difficult and have been solved only for smooth, spherical particles and for simple obstacles shapes such as spheres and cylinders. The mathematics are given by Brun and Mergler (1953) and by Green and Lane (1964).

Although theory deals only with efficiency of impaction, the efficiency of retention is also important. A particle, upon impact, may either stick to the obstacle or rebound from it and re-enter the air stream. A sampling surface must be coated with a good adhesive to insure adequate retention. Sampling efficiency is a product of impaction efficiency and retention efficiency and can be determined experimentally in a wind tunnel.

Since impaction efficiency is a function of wind speed, particle characteristics, and collector size; acceptable efficiency can be obtained only for certain combinations of these variables. Particles may impact on obstacles of any shape, but vertical cylinders are most commonly used as impaction samplers since they are horizontally symmetrical and their impaction efficiency can be calculated. The relationship between efficiency of impaction and cylinder size is illustrated in figure 5-2 and given by the equation

$$E = d/D$$

where

E = efficiency of impaction
D = cylinder diameter and
d = crosswind diameter from which particles impact.

Note that the ratio d/D is larger for the smaller cylinder; indicating that a smaller collector is more efficient than a large one, all other variables being equal. Figure 5-3 illustrates the change in efficiency with wind speed of two cylinder sizes for two representative pollens. Efficiency is negligible at very low wind speeds for both cylinders and both particles. The efficiency of the smaller cylinder increases more rapidly as wind speeds increase and always remains higher. The

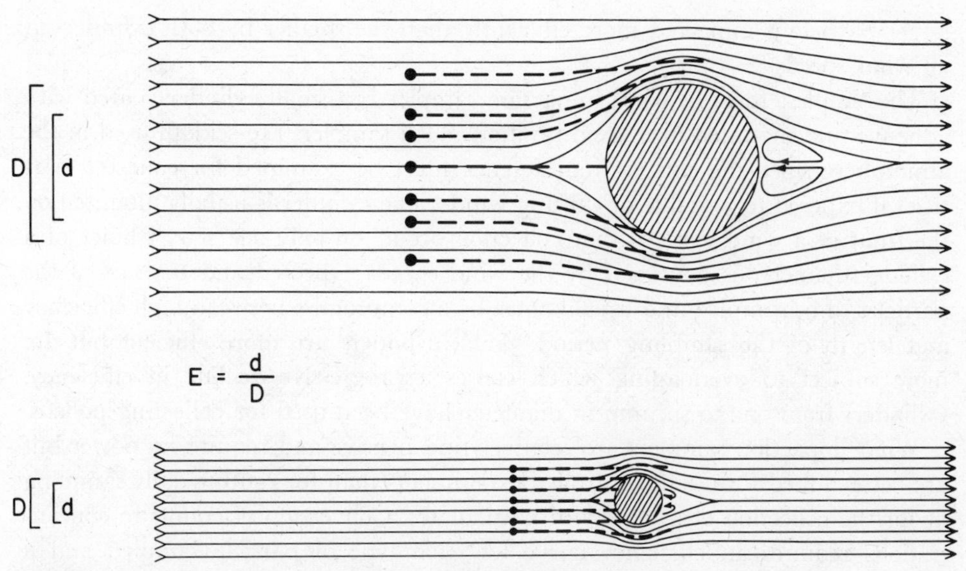

FIGURE 5-2 Air (solid lines) and particle (dashed lines) trajectories around large and small cylinders illustrating the greater impaction efficiency (E) of the small cylinder. E = d/D where d = partial diameter of the cylinder from which particles impact and D = cylinder diameter.

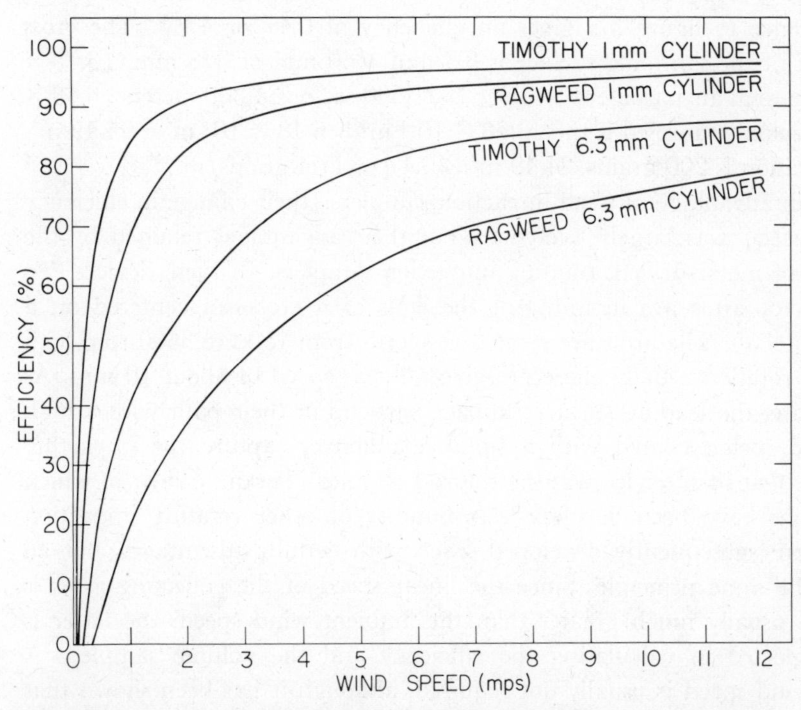

FIGURE 5-3 Computed curves relating efficiency of impaction of pollen particles of two diameters, ragweed (20 microns) and timothy (33 microns), on one-mm- and 6.3-mm-diameter cylinders as a function of wind speed.

larger particle is impacted more efficiently than the smaller by both cylinders at all wind speeds.

The simplest form of wind impaction sampler is a small cylinder coated with adhesive and mounted in a fixed position. Such samplers are seldom used in the atmosphere since the entire circumference must be examined for collected particles if exposed to variable or shifting winds. The cylinder is usually mounted on the front of a wind vane so that collection occurs on only one side. Choice of a cylinder diameter is governed by the wind speeds expected and the size of the particles to be sampled and usually must be a compromise between high efficiency and length of the sampling period. Small cylinders are more efficient but are more subject to overloading which causes a progressive decline in efficiency. Cylinders from one to six mm in diameter have been used for collecting pollens.

Wind impaction samplers are relatively inexpensive and require no power but are better suited for use in controlled experiments than for routine daily sampling or for the collection of a variety of particulates. Conversion of counts to concentrations requires an efficiency curve for each type of particle sampled and a measure of wind speed, preferably from a sensitive anemometer mounted close to the sampler. The mean wind speed is used to determine the sampling efficiency from the curve. The total air passage, found by multiplying the mean wind speed by time, is multipled by the cross sectional area of the projected sampling surface to give the volume of air sampled. The count divided by the volume times the efficiency gives the concentration.

As an example, suppose a 6.3-mm-diameter cylinder, 60 mm long, is exposed for 6 hours at a mean wind speed of 3 m/sec and collects 200 ragweed pollen grains. Reference to figure 5-3 gives an efficiency of 0.47 or 47%. The cross sectional area of the sampling surface is 6.3 mm $\times$ 60 mm or 378 mm^2 (3.78×10^{-4} m^2). The total air passage is 3 m/sec $\times$ 21,600 sec or 64,800 meters (6.48×10^4m). The volume sampled is then 3.78×10^{-4} m^2 $\times$ 6.48×10^4 m or 24.49 m^3. The concentration is 200 grains/24.49 m^3 $\times$ 0.47 or 17.4 grains/m^3.

The basic disadvantage of wind impaction samplers (their change in efficiency with wind speed) was largely overcome, but their advantages retained by the development of motor-driven, rotating impaction samplers. In these devices two vertical collector arms are mounted at the ends of a crossarm centered on a vertical motor shaft. The arms are rotated at speeds from 1500 to 3600 rpm with the radius of rotation usually chosen to give a linear speed of about 10 mps. As the arms rotate, the leading surfaces impact particles in their path with a high efficiency and, being coated with a suitable adhesive, capture the impacting particles. The first such device was the rotorod sampler (Perkins 1957) of which several versions have been developed. A number of other rotating impaction samplers were subsequently developed, each with certain advantages, but all operate on the same principle. Since the linear speed of the collecting arms is constant and usually much greater than the ambient wind speed, the latter is normally neglected in calculating the efficiency and the volume sampled. A measure of wind speed is usually not required, although it has been shown that efficiency decreases somewhat as wind speed increases (Ogden and Raynor 1967). The collection efficiency also varies with collector and particle size as in wind impaction samplers, but is normally much higher and can be calculated or measured experimentally in a wind tunnel.

If operated for prolonged periods, the high collection efficiency of these devices leads to overloading and versions have been designed which operate sequentially or intermittently. Some of these instruments actually rotate for only a small percentage of the total exposure time, and it is necessary to shield the collecting surfaces from wind impaction during their idle periods.

Rotating impaction samplers are well-suited for sampling airborne pollens, and their use by allergists and public health workers is increasing. Although they require power and may have occasional mechanical or electrical problems, they measure pollen concentrations with acceptable accuracy and their samples are nearly as easy to count as those from gravity slide samplers.

SUCTION

Samplers in which air containing material to be sampled is drawn into an entrance by suction from a vacuum pump or other air moving device may be classified as suction samplers and are used for many air sampling purposes. Many methods are used within such samplers for collecting the material of interest from the air stream. These methods include filtration, impaction, electrostatic and thermal precipitation, and liquid impingement. Some suction samplers do not collect the material but measure its concentration, usually by optical methods, as it passes through a special viewing or measurement section.

Techniques and devices for measuring and removing materials drawn into suction samplers have been subjects of extensive research and experimentation and are generally highly efficient for the specific type or size of particle for which the sampler is designed, but little attention has been given to the problem of getting a representative sample into the entrance. This is partially due to the fact that suction samplers are most commonly used to sample gases and submicron-sized particles whose entrance efficiency is normally high. However, such samplers are sometimes recommended and often used to sample larger particles, such as pollen, which tend to deviate from the air stream entering the sampler if the air is forced to change direction or speed. In such cases, the number collected may be much different (smaller or larger) from the number originally in the air sampled.

Isokinetic sampling (Watson 1954) is the ideal method of taking an accurate sample of large particulates, such as pollen, from the atmosphere. In this method, illustrated by the central diagram of figure 5-4, air is drawn into a sharp-edged orifice aligned with the air stream. Air within the sampler is drawn away from the entrance at the same velocity (V_S) as ambient air approaches (V_A). The approaching air does not have to change direction or speed and enters the sampler entrance smoothly carrying all entrained particulates with it. If V_A is greater than V_S, as in the upper diagram, not all the approaching air can enter and some must divert around the entrance. Particles in the diverted air may have enough inertia to carry them into the entrance and cause oversampling. If V_A is less than V_S, as in the lower diagram, air is drawn in from outside the stream approaching the entrance, but particles in this region may have enough inertia to carry them past the entrance and undersampling results. Isokinetic sampling is a standard method of obtaining an accurate sample in wind tunnels, tubes, ducts, and chimneys where the flow is constant, but samplers are not yet perfected which can

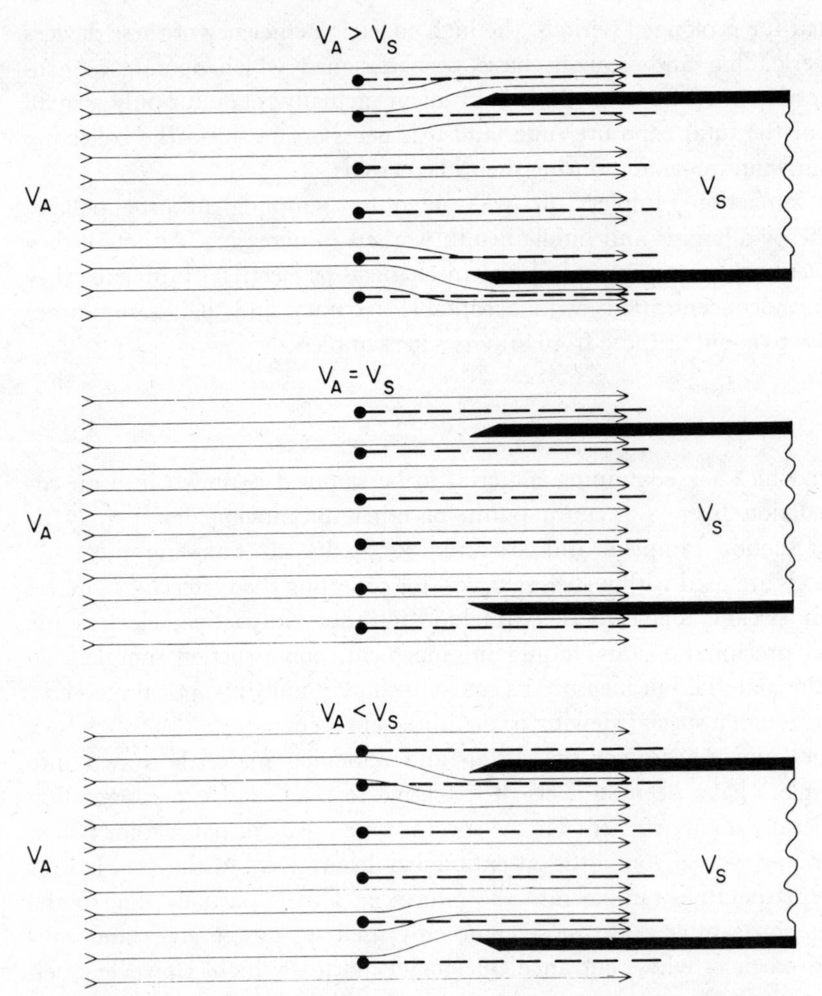

FIGURE 5-4 Air (solid lines) and particle (dashed lines) trajectories at the entrance of a suction type sampler facing the wind. Upper diagram—air speed (V_A) exceeds sampling speed (V_S). Particles in air which diverts around entrance enter resulting in oversampling. Middle diagram—$V_A = V_S$. All particles in air sampled enter sampler (isokinetic sampling). Lower diagram—V_S exceeds V_A. Particles in air which enter sampler escape resulting in undersampling.

adjust to the rapid fluctuations in speed and direction present in the free atmosphere. If samplers are vane-mounted and have a flow rate which is adjustable, they can be used to approximate isokinetic sampling by matching the flow rate to the average wind speed but most samplers are designed for a fixed flow rate only. Equations for estimation of the errors involved in nonisokinetic sampling are given by Badzioch (1960).

Most suction samplers are so designed that even an approximation to isokinetic sampling is impossible. In samplers where the entrance does not face the air stream, flow patterns near the entrance are complex and entrance efficiency for large particles is generally low. Air and particle trajectories around a typical filter sampler with the entrance at a right angle to the air stream are illustrated in

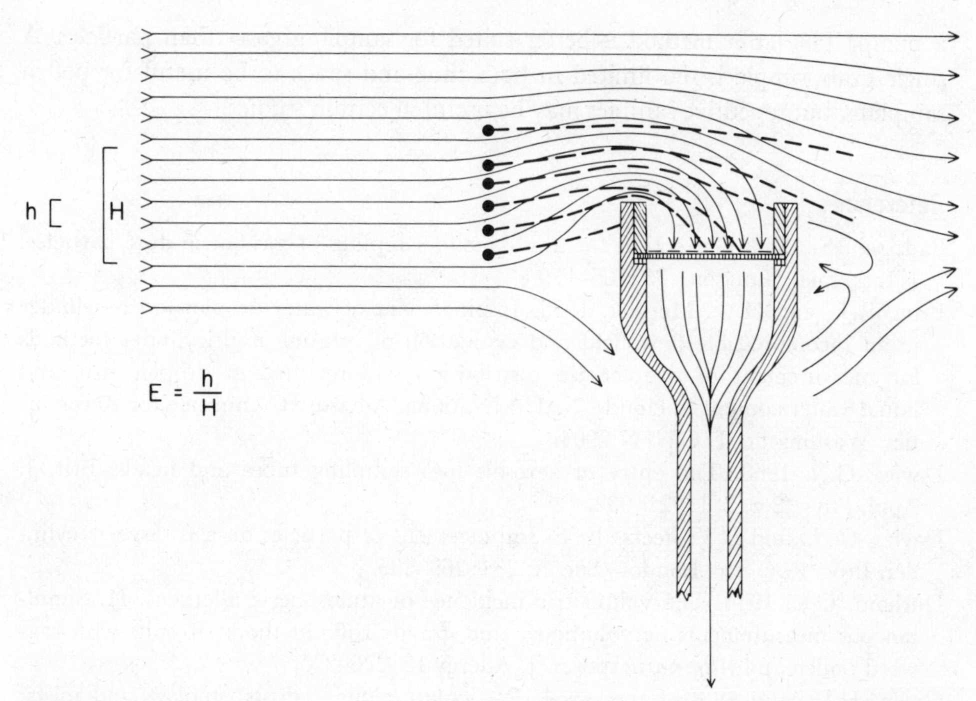

FIGURE 5-5 Air (solid lines) and particle (dashed lines) trajectories around a filter sampler at right angles to the air flow. Entrance efficiency (E) is given by h/H where h is the height of the air column from which particles enter the sampler and H is the height of the air column entering the sampler.

figure 5-5. Many of the particles originally in the air which enters the sampler pass by or strike the filter holder and are not captured. The efficiency (E) is given by h/H where h = extent of the vertical air column from which particles enter the sampler and H = extent of the vertical air column which enters the sampler. Wind tunnel experiments and field experience with filter samplers, oriented as in figure 5-5, show that entrance efficiency increases with increased flow rate into the entrance and decreases with increasing wind speed and with increasing particle size. It may drop to near zero for pollen-sized particles at wind speeds often found in the atmosphere and with flow rates commonly used. These results suggest that suction samplers are not normally suitable for sampling pollens. Although they may be 100% efficient in removing and retaining particles which enter them, their low and variable entrance efficiency for large particles is seldom mentioned by the manufacturer and is not known to many users.

GRAB SAMPLING

Grab sampling consists of quickly capturing a volume of air by some means, hopefully without changing the concentration of particles contained therein. The sample is normally returned to a laboratory where the particles or other constituents of interest are removed from the air and counted or measured in some way. The method may be illustrated by suddenly covering a container such as a can or cylinder open at both ends to the free flow of air. Methods used in the past for grab sampling included opening a previously evacuated container allowing the air to rush in or quickly blowing up a balloon or plastic bag with a pump. The latter method is better suited for sampling gases than particles. A single grab sample is too limited in both time and space to be useful for pollen sampling, but repetitive samples may be useful in certain studies.

References

Badzioch, S. 1960. Correction for anisokinetic sampling of gas-borne dust particles. J. Inst. Fuel (London) 33:106–110.

Brun, R.J., and H.W. Mergler. 1953. Impingement of water droplets on a cylinder in an incompressible flow field and evaluation of rotating multicylinder methods for measurement of droplet-size distribution, volume-median droplet size, and liquid-water content in clouds. NACA [National Advisory Committee for Aeronautics, Washington, D.C.] TN 2904.

Davies, C.N. 1968. The entry of aerosols into sampling tubes and heads. Brit. J. Appl. Phys., Ser. 2, 1:921–932.

Davies, C.N., and C.V. Peetz. 1956. Impingement of particles on a transverse cylinder. Proc. Roy. Soc. London, Ser. A, 234:269–295.

Durham. O.C. 1944. The volumetric incidence of atmospheric allergens. II. Simultaneous measurements by volumetric and gravity slide methods. Results with ragweed pollen and *Alternaria* spores. J. Allergy 15:226–235.

Green, H.L., and W.R. Lane. 1964. Particulate clouds: dusts, smokes, and mists; their physics and physical chemistry and industrial and environmental aspects. 2nd ed. D. Van Nostrand, Princeton, N.J.

Gregory, P.H. 1951. Deposition of air-borne *Lycopodium* spores on cylinders. Ann. Appl. Biol. 38:357–376.

Gregory, P.H. 1973. The microbiology of the atmosphere. 2nd rev. ed. John Wiley & Sons, New York. *See* especially Chapters 2, 7, 8, and 9.

Gregory, P.H., and O.J. Stedman. 1953. Deposition of air-borne *Lycopodium* spores on plane surfaces. Ann. Appl. Biol. 40:651–674.

Lewis, D.M., and E.C. Ogden. 1965. Trapping methods for modern pollen rain studies. Pages 613–626 *in* B. Kummel and D. Raup, eds. Handbook of paleontological techniques. W.H. Freeman & Co., San Francisco.

Ogden, E.C., and G.S. Raynor. 1960. Field evaluation of ragweed pollen samplers. J. Allergy 31:307–316.

Ogden, E.C., and G.S. Raynor. 1967. A new sampler for airborne pollen: the rotoslide. J. Allergy 40:1–11.

Perkins, W.A. 1957. The rotorod sampler. 2nd semiannual report, CML. 186. Aerosol Laboratory, Stanford Univ., Standford, Calif.

Ranz, W.E. 1956. Principles of inertial impaction. Pa. State Univ. Eng. Res. Bull. 66.

Raynor, G.S. 1970. Variation in entrance efficiency of a filter sampler with air speed, flow rate, angle and particle size. Amer. Ind. Hyg. Ass. J. 31:294–304.

Richardson, E.G., ed. 1960. The aerodynamic capture of particles. (Proc. of a conf. held at Leatherhead, England) Pergamon Press, New York. 200 p.

Solomon, W.R. 1967. Techniques of air sampling. Pages 326–339 *in* J.M. Sheldon, R.G. Lovell, and K.P. Mathews. A manual of clinical allergy. 2nd ed. W.B. Saunders Co., Philadelphia.

Vitols, V. 1966. Theoretical limits of errors due to anisokinetic sampling of particulate matter. J. Air Pollut. Contr. Ass. 16:79–84.

Watson, H.H. 1954. Errors due to anisokinetic sampling of aerosols. Amer. Ind. Hyg. Ass. Quart. 15:21–25.

6
POLLEN SAMPLERS

Many sampling devices operating on the principles described in the preceding Chapter have been used for sampling pollens and other airborne particles. These range from simple, time-honored devices with no moving parts to highly sophisticated instruments incorporating the latest advances in electronics, optics, and fluid mechanics. Some are designed for specialized tasks; others have more general usefulness. Some can be used only for a limited range of particle sizes. Many of these were designed for sampling submicron-sized particles while few were designed for particles as large as pollen. Some were developed for sampling indoor air as in clean rooms; others are built for exposure to the elements. Only those samplers that have been or could be used for sampling airborne pollens with reasonable success will be discussed in this Chapter. Such samplers should have as many as possible of the following characteristics:

1. The sampler should have a reasonably high efficiency for the particles of interest under all normal operating conditions.
2. If the efficiency varies with wind speed or other factors, the manner of variation should be known.
3. The sampler should sample a large enough volume of air to give a representative sample even when concentrations are low.
4. The efficiency should not change significantly due to overloading before the sampler has operated for a long enough period or has taken a large enough volume of air to give a satisfactory sample.
5. The volume of air sampled per unit time preferably should be constant; but, if not, it should be capable of being calculated simply and accurately from associated data such as the wind speed.
6. Changing, storing, and examination of the samples should be reasonably simple.
7. The sampler should be so designed and constructed that it will not be damaged by exposure to the normal range of atmospheric conditions.
8. The sampler should be commercially available, or easily constructed, at a cost low enough to permit general use.

FIGURE 6-1 Durham or gravity slide sampler. A greased microscope slide is held in the horizontal support between the two protective plates.

GRAVITATIONAL SETTLING SAMPLERS

This category includes all sampling devices in which the sampling surface is exposed in a horizontal position either on the ground or at some elevation. As explained in the previous Chapter, capture takes place by turbulent impingement as well as by gravitational settling. Retention of settled particles is not normally a problem with adhesives commonly used except that rain or heavy dew may loosen some particles and float them away. Wind, of course, would remove many settled particles if no adhesive were used.

Durham sampler

The Durham sampler (Fig. 6-1) consists of a mount for positioning a glass microscope slide between two horizontal 9-inch circular disks. It is usually mounted on a metal rod or pipe support at least several feet above the ground or on a roof top. It was designed by Oren C. Durham, for many years Chief Botanist at Abbott Laboratories, and was adopted as the standard pollen sampler by the Pollen and Mold Committee of the American Academy of Allergy (Durham 1946). It is commercially available and is still widely used by allergists, hospitals, and public health agencies. Pollen collection is by gravitational settling and by turbulent impingement to the greased surface of a microscope slide.

Advantages of the Durham sampler are:

1. The slides are easily loaded and counted.
2. It is inexpensive.
3. It has no moving parts.
4. It requires no electric power.

Disadvantages are:

1. The volume of air sampled is unknown, so the catch cannot be converted to a volumetric measure of concentration.
2. The efficiency cannot be determined.

3. The catch is relatively low.
4. The catch is a function of wind speed, turbulence, and orientation of the sampler with respect to wind direction as well as concentration of pollen in the air.

These disadvantages have recently been demonstrated experimentally (Ogden and Raynor 1960), but most were realized by previous workers, including Mr. Durham. Several attempts have been made to improve the sampler. A square slide mounted in a streamlined, circular holder has been used to eliminate the directional effect. Some workers have mounted slide samplers on vanes and some have tilted them at a 45-degree angle in attempts to collect by both impaction and settling. None of these modifications appears to be a significant improvement over the original sampler.

Counts from the Durham sampler are so greatly influenced by factors other than the concentration of pollen in the air that they indicate only in a general way the presence and abundance of pollen in the atmosphere. Tests have shown that it is impossible to determine the volume of air sampled; therefore, the data are qualitative at best (Dingle 1957, Harrington et al. 1959, Ogden and Raynor 1960, Hayes 1969). Daily counts are highly misleading. Counts from different localities are not properly comparable unless the influencing meteorological parameters are the same. However, averages of the daily counts over a pollination season are useful for comparison with other localities as the influencing factors usually tend to average out.

The Durham sampler has served a useful purpose as it has encouraged the sampling for ragweed and other pollens in hundreds of localities in North America over a period of many years. These figures, being averaged for each locality, allow comparisons which, in the absence of any other data, are useful. The daily counts, as taken with the Durham sampler and reported by the news media, should be interpreted with an understanding of the shortcomings of the sampler. This the allergist has learned to do, but the layman has not. More unfortunately, counts are reported from samplers which are not installed and operated according to the procedures suggested by Durham and adopted by the American Academy of Allergy.

Too often the "pollen counts" are obtained under conditions that make them worthless, highly misleading, and a disservice to all concerned. Because of these factors, use of this sampler should be discouraged and a sampler capable of making reasonably accurate, quantitative measurements adopted.

Deposition samplers

Measurements of pollen deposition to the earth's surface are useful in some studies and may be taken by several methods. The simplest is the use of microscope slides coated with an adhesive and laid on the ground. All or a known portion of the slide may be counted and the results expressed as grains per area, usually grains/m^2. Sheets of sticky paper in a flat aluminum frame (Fig. 6-2) as used by the U. S. Atomic Energy Commission to collect radioactive fallout particles (Rosinski 1957) are also good collectors and retain collected particles better during wet weather than other commonly used adhesives (Lewis and Ogden 1965). Selected portions may be cut from the sheet for examination under the micro-

FIGURE 6-2 AEC fallout paper in flat aluminum frame for measuring deposition to the ground.

scope. These surfaces will not necessarily collect the same number of particles per unit area as would the normally rougher surrounding soil or vegetated surface.

Petri dishes with sugar to simulate the sand surface in which the dishes were embedded have been used. Deposited pollen is removed from the liquid in which the sugar has been dissolved. Pans of water or other liquid may also be used as deposition collectors. Pollen may be removed and concentrated by filtration of the liquid. These methods can be recommended only for periods of dry weather. Attempts to compute air concentrations from deposition data should not be made unless the velocity of deposition to the sampling surface has been established for a particular pollen.

WIND IMPACTION SAMPLERS

The use of impaction samplers for pollen sampling is a fairly recent development; the types described below were all introduced within the past fifteen years. They seem likely to supplant other types since they are inherently more suitable for sampling particles as large as pollen. As discussed in the previous Chapter, these devices may be divided into wind impaction and powered impaction samplers, both of which collect on surfaces at right angles to the wind.

Only two wind impaction samplers have been used extensively in the field and both of them in large-scale pollen research projects.

Flag sampler

The flag sampler (Fig. 6-3) consists of a two-inch length of one-inch-wide transparent cellulose tape wrapped around a two-inch-long straight pin, one mm in diameter (Harrington et al. 1959). The tape is pressed together except near the tail where the ends are separated by a thin piece of folded paper to facilitate removal after exposure. The portion of tape around the pin is coated with an adhesive. The pin is inserted in a ¾-inch length of ⅛-inch glass tubing sealed at the bottom. This serves as a bearing and allows the flag to move freely with the wind. After exposure, the tape is removed and mounted on a microscope slide

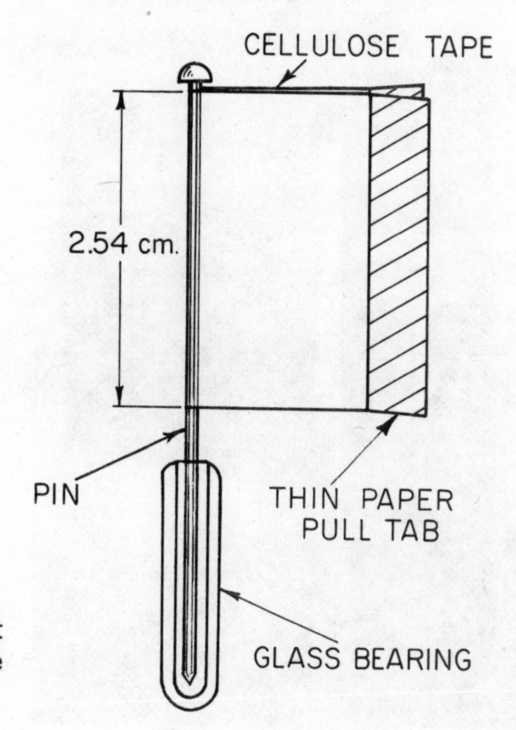

FIGURE 6-3 Flag sampler. Particles impact on the adhesive-coated edge of the tape wrapped around the pin.

for examination. This sampler is a modification, by workers at the University of Michigan, of the Sakagami flag wind vane.

Advantages of the flag sampler are:

1. It is inexpensive.
2. It has no moving parts.
3. It does not require power.
4. It is easily constructed.
5. Its efficiency can be computed if characteristics are known or determined experimentally for particles of interest.
6. The volume sampled can be determined if the wind speed is known.

Disadvantages are:

1. The efficiency and the volume sampled vary with wind speed; so if quantitative measurements are desired, the wind speed must be measured.
2. The efficiency varies with particle size and density.
3. The small size may lead to overloading during prolonged sampling periods if concentrations of airborne particulates are high.
4. Difficulty is sometimes experienced in separating the two halves of the tape and in placing the exposed portion flat on a slide.
5. It is not commercially available.

Slide-edge-cylinder sampler

The slide-edge-cylinder sampler (Fig. 6-4) was developed at Brookhaven National Laboratory (Raynor et al. 1970). The leading edge of the sampler consists of two aluminum bars, 4.5 in. long, ¾ in. wide and ⅛ in. thick, machined

FIGURE 6-4 Slide-edge-cylinder sampler. Particles impact on the adhesive-coated edge of a microscope slide inserted in the spring-loaded bars.

to a semicylindrical shape in front. A microscope slide placed between the bars is held in place by spring tension with its leading edge flush with the metal bars. This slide edge is coated with an adhesive and serves as the sampling surface. The sampler has a splayed vane tail and rotates on a pointed shaft inserted in the cylindrical tube near the center. The tip of the shaft engages the concave bottom of an allen set screw in the top of the tube. A curve of efficiency of impaction for ragweed pollen is shown in figure 6-5.

Advantages of the slide-edge-cylinder are:

1. It is relatively inexpensive.
2. It has no moving parts.
3. It does not require power.
4. Its efficiency can be computed if particle characteristics are known or it may be determined experimentally for particles of interest.
5. The volume sampled can be determined if the wind speed is known.
6. The sample is taken on the edge of a microscope slide which is easy to insert in the sampler, store for later analysis, and examine under the microscope.

Disadvantages are:

1. The efficiency and the volume sampled vary with wind speed; so if quantitative measurements are desired, the wind speed must be measured.
2. The efficiency varies with particle size and density.
3. The efficiency is low at low wind speeds.
4. It is not commercially available.
5. A slide positioner (Fig. 8-4) is necessary to hold the slide on edge under the microscope.

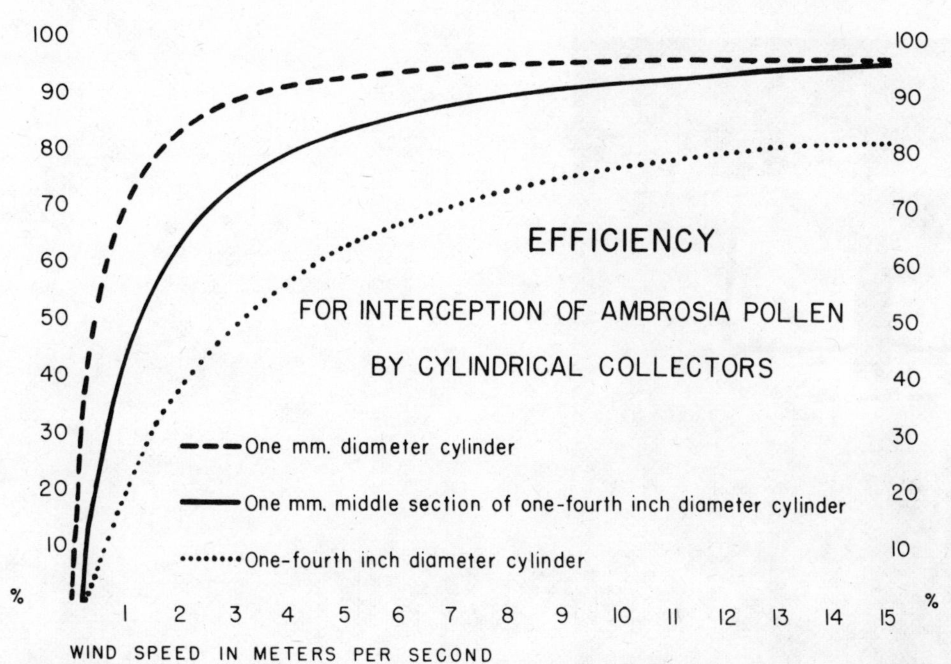

FIGURE 6-5 Graph showing calculated efficiency of impaction with wind speed on 1-mm and 1/4-inch cylinders and on the sampling surface of the slide-edge-cylinder sampler.

Since the flag sampler and the slide-edge-cylinder sampler are identical in principle, a choice between the two should be made on the basis of efficiency needed, length of sampling period, and type of sampling surface. The slide-edge-cylinder is considered to be preferable for most situations because of the ease of using the microscope slide. Neither is recommended for general, routine use because of the variation in efficiency with wind speed and the need for an accompanying anemometer.

Slide-edge-free sampler

A wind impaction sampler can also be constructed by inserting a microscope slide in a holder on the front of a wind vane. In this "slide-edge-free" sampler, the leading narrow edge of the slide is not flush with its support but protrudes for some distance. Its characteristics are similar to those of the slide-edge-cylinder. The efficiency is somewhat higher since the width of the slide alone is less than the width of the slide and supporting bars in the slide-edge-cylinder. However, alignment with the wind is more critical in the slide-edge-free sampler since imperfect alignment causes part of the sample to deposit on the side or fail to deposit at all.

ROTATING IMPACTION SAMPLERS

Several rotating impaction samplers are in use.

Rotorod sampler

The first developed was the rotorod sampler (Perkins 1957). Figure 6-6 illustrates a rotorod built at Brookhaven National Laboratory which is identical in all im-

FIGURE 6-6 Rotorod sampler. The horizontal section of the sampling arms is bent to fit over the grooved hub on the shaft of a DC motor in the can. The arms impact particles on their forward sides as they rotate.

portant aspects to the original. The sample is collected on the upright $\frac{1}{16}$-inch-square metal rods which are coated with an adhesive on the leading edges. The unit is powered by a small battery-operated DC motor in a protective case. It rotates at 2500 rpm giving a linear speed of 10.6 mps and samples 120 liters/min. The arms are 6 cm long and 4 cm from the center of rotation. Both arms are made from one piece of metal designed to slip over a special hub on the motor shaft. This sampler has a high efficiency for pollen-sized particles, but the arms are troublesome to handle without disturbing the sample and are difficult to place under a microscope. The sample can be viewed only in reflected light.

A later version, the Rotorod FP Sampler (Webster 1963) uses H-shaped metal rods with much narrower (0.015-inch) collecting surfaces. This was designed for collecting fluorescent particle tracer material, only a few microns in size, and is not recommended for pollen sampling.

A model designed specifically for sampling larger particles and for microscope viewing of the sampling surface uses plastic inserts on the upright rods (Webster 1968). It has dimensions and specifications similar to the original rotorod sampler and is suitable for pollen sampling (Fig. 6-7). However, the small plastic inserts are somewhat difficult to grease evenly unless inserted in their holder or a special

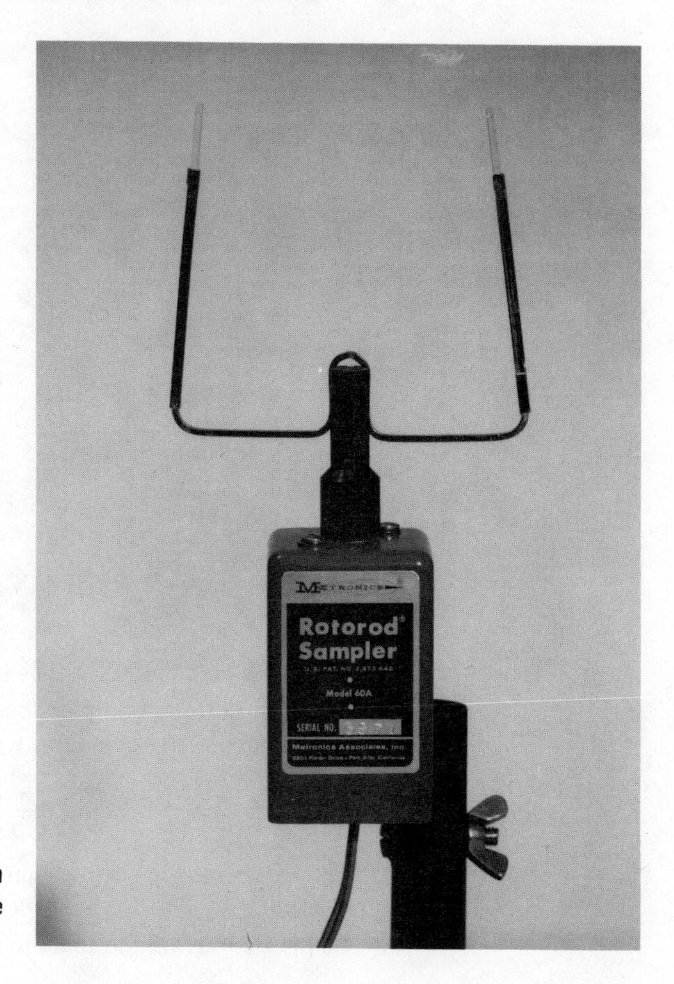

FIGURE 6-7 Rotorod sampler for pollen sampling. Plastic inserts mounted in the metal uprights are removable for counting.

fixture. After exposure, they may be taped to a microscope slide for examination or inserted in grooves in a small piece of transparent plastic.

None of these versions has provision for shielding the collecting surfaces when not in use, so they are useful only for continuous operation. In a recent version designed for intermittent use (Metronics Associates 1967), the collector arms, which hang down from a crossarm mounted on the motorshaft, retract into protective slots in the crossarms when not rotating. Centrifugal action forces the arms into sampling position during rotation (Fig. 6-8).

Advantages of this version are:

1. It is suitable for intermittent and sequential as well as continuous use.
2. The volume sampled is known.
3. The efficiency is high and may be calculated or determined experimentally for specific particles.
4. It is battery-powered and can be operated easily at remote locations.
5. It is commercially available at moderate cost.

Disadvantages are:

1. The efficiency varies with particle size and density.

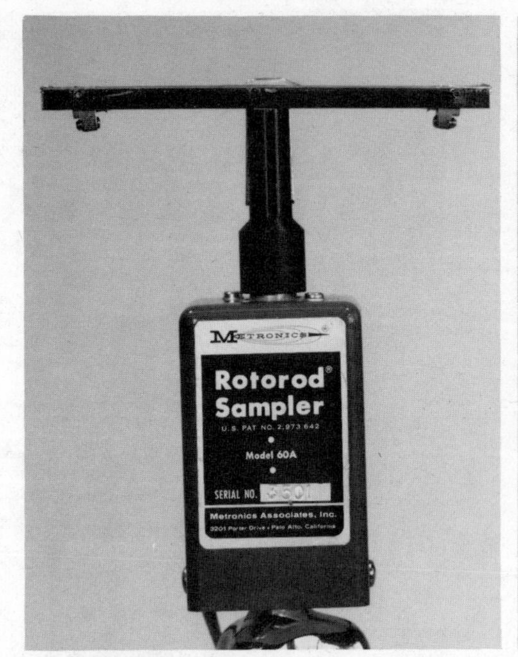

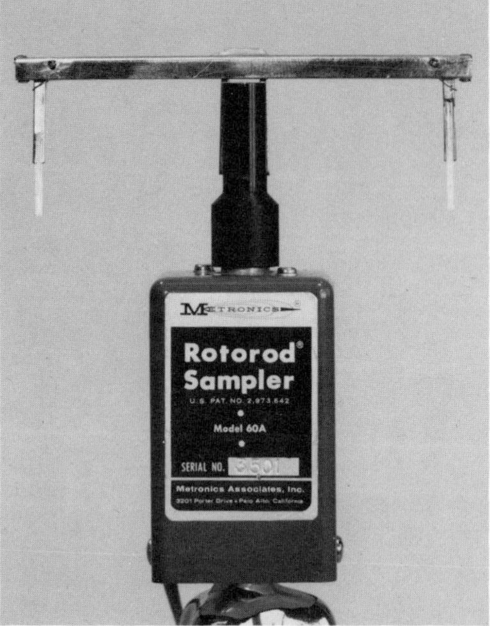

FIGURE 6-8 Rotorod sampler for intermittent sampling. The arms retract into protective slots in the crossarm when idle but extend vertically downward upon rotation.

2. The plastic rods are somewhat difficult to handle, store, and place under a microscope.
3. The plastic rods are too small to mark easily for identification.

Rotobar sampler

The rotobar sampler (Harrington et al. 1959) collects particles on two upright metal bars held in place by spring tension between two crossarms. The sampling arms are 38 mm long, 6 mm wide and 0.8 mm thick. The narrow edge serves as the sampling surface and is covered with adhesive tape (bringing the thickness to 1.0 mm) on which the sample is deposited. The sampler is rotated at 3600 rpm by an AC motor and samples nearly one m^3 per hour.

A recent modification, the fly-shield rotobar (Solomon et al. 1968) uses simple hinged shields which protect the sampling surface when idle and fly out to a horizontal position by centrifugal force during rotation (Fig. 6-9). The shielding system is simple and inexpensive, but the shields may be dislodged from their protective positions by a strong wind from certain angles. Rotobar samplers are suitable for pollen sampling, but they should be compared to the rotorod and rotoslide samplers.

Advantages of the fly-shield rotobar sampler are:

1. It is suitable for intermittent and sequential as well as continuous use.
2. The volume sampled is known.
3. The efficiency is high and may be calculated or determined experimentally for specific particles.

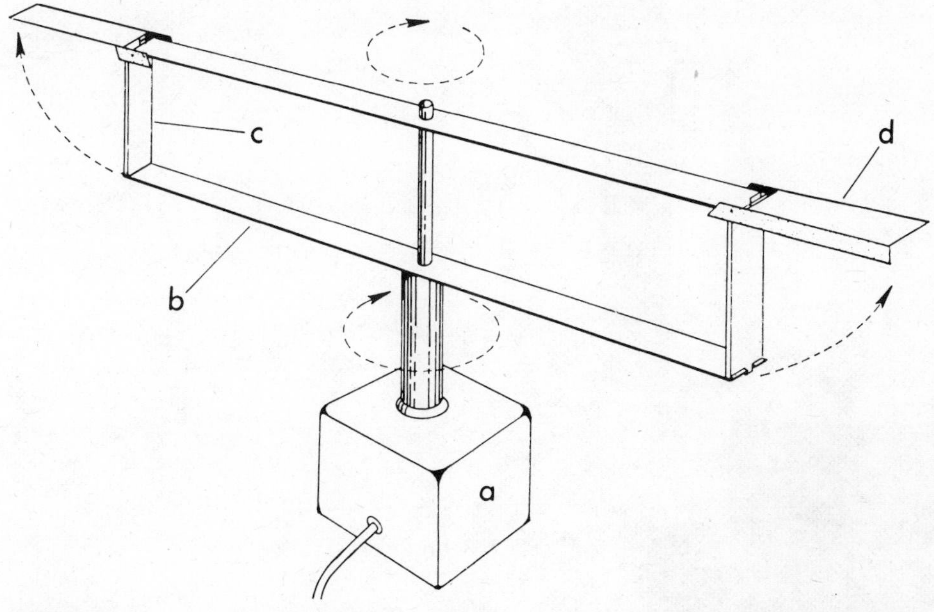

FIGURE 6-9 The fly-shield rotobar. The sampling bars (c) are held by flexible strips (b) and protected by the fly-shield (d) when not rotating. When rotated by the motor (a), the shields fly up exposing the adhesive-coated sampling surfaces.

4. The cost is moderate.

Disadvantages are:

1. The efficiency varies somewhat with particle size and density.
2. The tape on which the sample is taken is difficult to handle.
3. The protective shields can be dislodged from their protective position, when the sampler is idle, by wind from certain angles.
4. It is not available commercially.
5. Electric power is required.

A modified rotorod sampler (Magill et al. 1968) is similar to the nonshielded rotobar sampler. It consists of two upright bars 1.54 mm thick supported by upper and lower crossarms at 4 cm from the center of rotation. The arms rotate at 2400 rpm and sample 80 liters/min. The sample is transferred to the surface of a microscope slide for examination.

Rotoslide sampler

The rotoslide sampler (Fig. 6-10) consists of two upright slide holders at the ends of a horizontal crossarm mounted on the upright shaft of an AC motor. The slide holders are angled with respect to the crossarm so that the leading edge of each slide remains tangent to the circle while rotating. The radius of rotation is 5.7 cm, rotational speed about 1650 rpm, and linear speed about 10 mps. The volume sampled by a 50-mm length on both of the slides is about 0.059 m^3/minute or 3.54 m^3/hour. Due to the high collection efficiency of rotating impaction samplers,

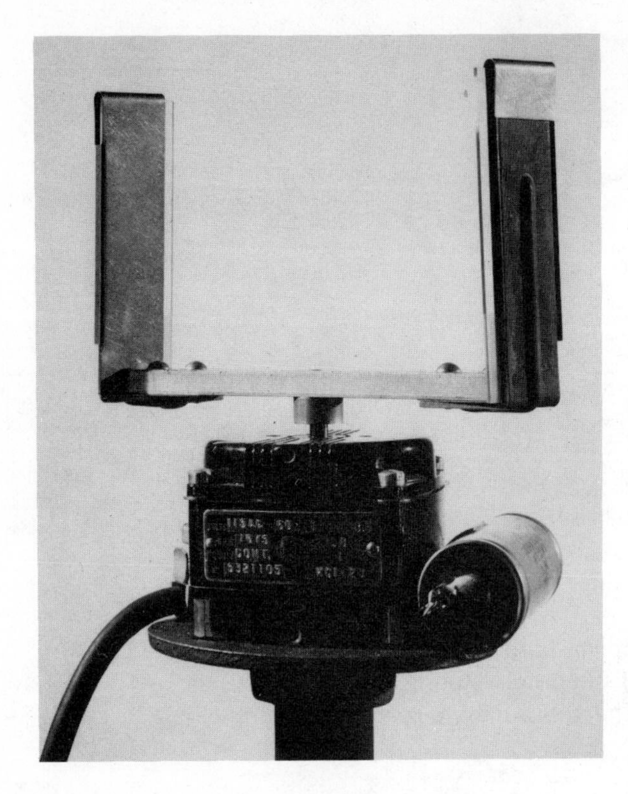

FIGURE 6-10 The rotoslide sampler. Microscope slides are inserted in the upright holders. The exposed edge collects particles upon rotation on an adhesive coating.

continuous operation for more than 2–3 hours may result in overloading unless concentrations of airborne particles are low.

Advantages of the rotoslide sampler are:

1. The volume sampled is known.
2. The efficiency is high and may be calculated or determined experimentally for specific particles. The efficiency has been determined for ragweed pollen.
3. The slides are easy to insert, store, and examine under the microscope.
4. The cost is moderate.
5. It is commercially available.

Disadvantages are:

1. It may overload during prolonged sampling periods.
2. The efficiency varies somewhat with particle size and density.
3. It requires electric power.
4. It is subject to wind impaction if exposed while idle.
5. A slide positioner (Fig. 8-4) is necessary to hold the slide on edge under the microscope.

The intermittent rotoslide sampler (Ogden and Raynor 1967) is designed for sampling over time periods as great as 24 hours (Fig. 6-11). It includes a timing device which actuates the sampling motor for one minute of each twelve giving a two-hour sample in 24 hours. A motor-driven cylindrical hood rises automatically to shield the slides from wind impaction when idle and lowers during rotation. A rain shield protects the instrument during precipitation.

The advantages are identical to those listed above. In addition, this version is suitable for long-period sampling. The disadvantages are also similar except that overloading and exposure to wind impaction are eliminated. Also, minor mechanical difficulties have been experienced with the cord and pulley system which raises and lowers the hood.

A more recent version, the swing-shield rotoslide (Raynor and Ogden 1970) uses a simplified shielding method. The curved shields protect the slide edges from wind impaction when not rotating, but are drawn away by centrifugal force and air pressure on the tails upon rotation to expose the sampling surfaces. (Figs. 6-12 and 6-13).

The advantages and disadvantages are similar to those listed above for the intermittent rotoslide except that the mechanical difficulties with the hood-raising mechanism are eliminated. The efficiency is essentially identical to the earlier versions. In addition, this version is well-adapted for sequential sampling by using an appropriate number of rotoslides with a suitable timing device (Fig. 6-14).

All rotoslide models described are available commercially. We have used the simple model and the intermittent model in quantity successfully over a period of years. The intermittent rotoslide has also been used by many allergists, public health departments, and universities. The swing-shield rotoslide has had four years of successful operation in the field. It is recommended as the most useful rotating impaction-sampler for use over lengthy periods.

With all rotating impaction samplers, it is important that the spin motors be dependable and that the speed of rotation be reasonably constant. The rpm at a given voltage will be supplied by the manufacturer of the sampler or it may be

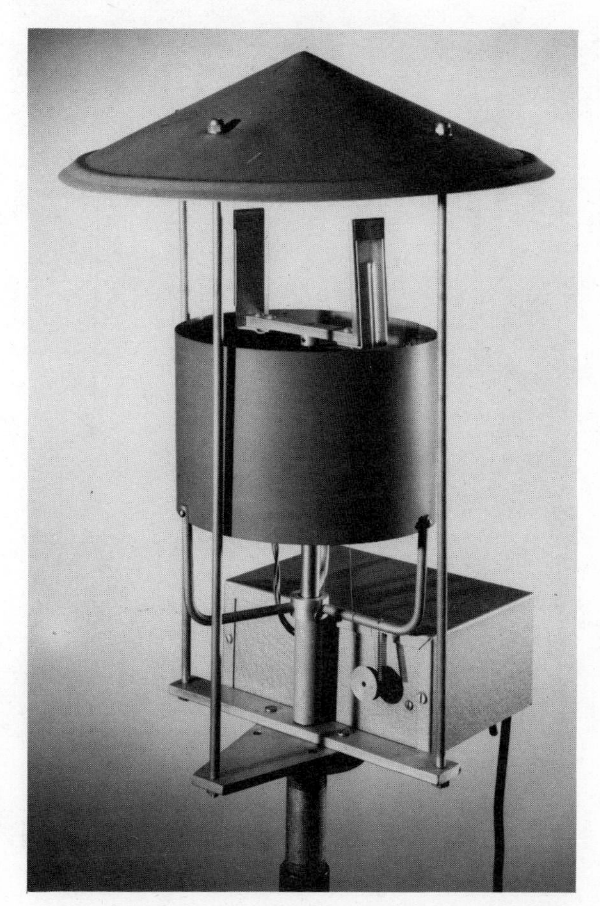

FIGURE 6-11 The intermittent rotoslide sampler. The cylindrical hood is pulled up by a motor and pulley system and protects the adhesive-coated slide edge from wind impaction when not rotating. Upon rotation, the hood automatically descends to the position shown.

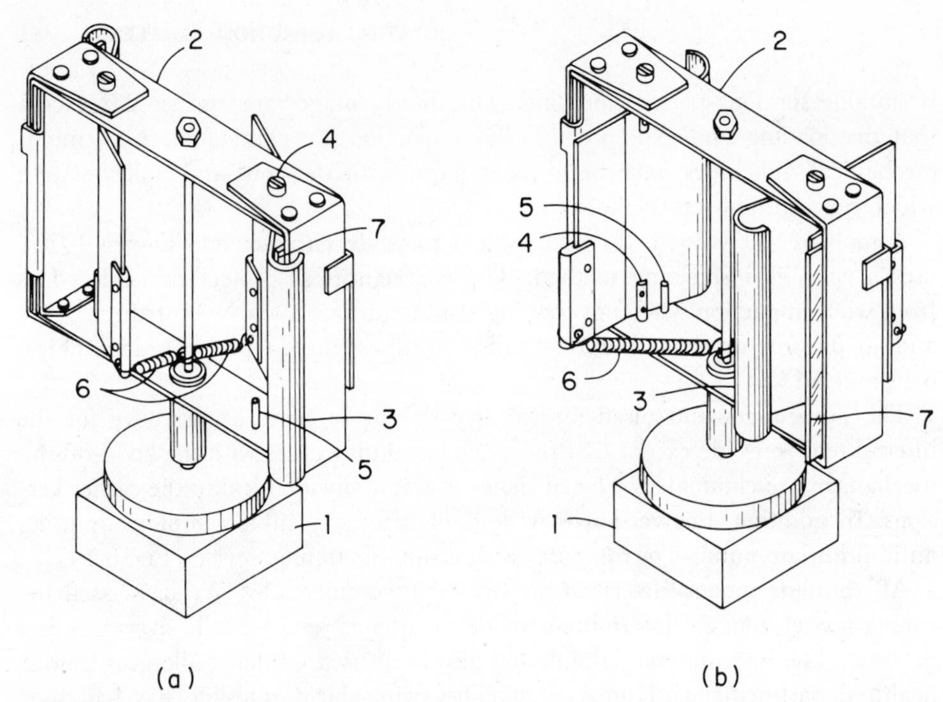

FIGURE 6-12 The swing-shield rotoslide. In the nonoperating position (a), the curved shields protect the sampling surfaces. During rotation (b), the shields are drawn away by centrifugal force and air pressure on the tail exposing the slide edges. Components include: (1) motor, (2) frame, (3) shielding plate, (4) pivot pins, (5) stop pins, (6) springs, and (7) glass slide.

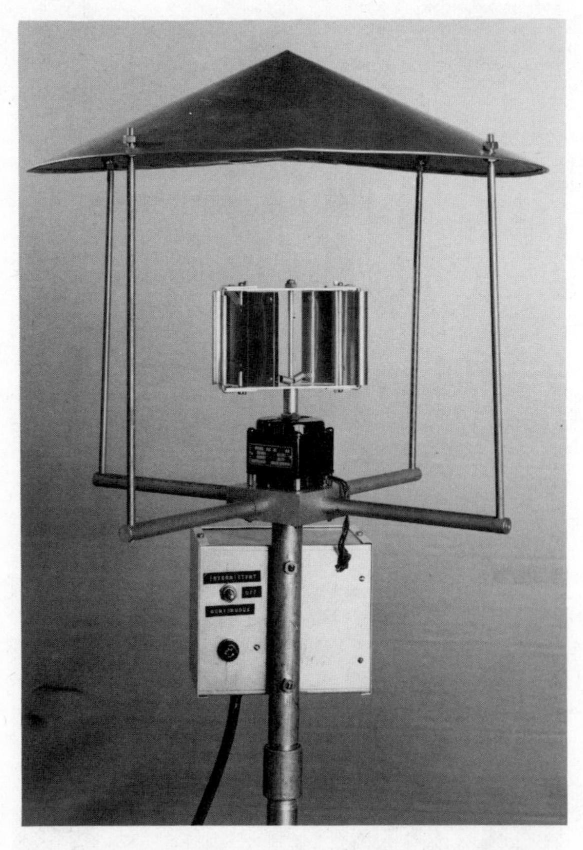

FIGURE 6-13 The intermittent swing-shield rotoslide sampler.

FIGURE 6-14 An assemblage of swing-shield rotoslide samplers with a timing device for sequential operation.

determined with a stroboscope. Generally, the speed of the rotoslide sampler will change about five rpm per volt, thus a drop in voltage of 10 volts would result in a drop of 50 rpm. Usually, this is of little importance. As an example, if a rotation of 1600 rpm at 120 V dropped to 1550 at 110 V, this is a reduction in the volume of air sampled of only about 3%. For situations where constancy of voltage may not be reliable, it may be advisable to attach a mechanical counter for recording the actual number of revolutions for each sample (Fig. 6-15).

Roto-disk pollen sampler

The sequential roto-disk pollen sampler (Cole and Stohrer 1966) takes sequential samples on discrete segments of the edge of a rotating disk by use of timing and stepping devices.

SUCTION SAMPLERS

The advantages and disadvantages of suction-type samplers for collecting pollen and other large particles were explained in Chapter 5. Several such devices are useful for certain studies.

Hirst spore trap

The Hirst spore trap (Hirst 1952) was the first useful suction-type sampler readily available for sampling pollens and other spores (Fig. 6-16). The vane tail keeps

FIGURE 6-15 Tetrabar rotoslides with counters. Left, with 110V AC motor. Right, with 12V DC motor.

FIGURE 6-16 The Hirst spore trap. Particles are sucked into the orifice beneath the rain shield and impacted on an adhesive-coated microscope slide which is pulled upward at 2 mm/hour by clockwork. The tail keeps the orifice facing the wind.

the 2 × 14 mm intake orifice facing the wind, and a rain shield protects the orifice from precipitation. The flow rate is 10 liters/minute which is monitored by a built-in flowmeter. It must be provided with an external vacuum pump. One with a one-sixth horsepower motor is adequate. The pump should have a gate valve on the intake line to set the flow through the sampler since the valve in the sampler allows the flow to fluctuate when exposed in the wind.

The efficiency, though variable with wind speed and with particle size, is reasonably high. Inside the housing containing the orifice, a greased microscope slide is drawn upward by clockwork at a rate of 2 mm/hour. Particles in the air sampled are deposited by impaction on the slide which is changed each day. Tests have shown that most of the entering pollen is deposited on the strip of slide directly behind the orifice, but that a small percentage spreads in both directions. This effect, plus the constant motion of the slide, results in a smoothed hourly average when the pollen grains on a portion of each 2 mm strip are counted.

Advantages of the Hirst spore trap are:

1. It measures variation in concentration with time.
2. The efficiency is reasonably high.
3. The slides are easy to insert, handle, and count.
4. It is commercially available at moderate cost.
5. This sampler is rugged, well-constructed, and not affected by exposure.

Disadvantages are:

1. The efficiency varies with wind speed and with particle size and density.
2. Electric power and an external vacuum pump are required.

Burkard recording volumetric spore trap

The Burkard seven-day recording volumetric spore trap (Burkard Manufacturing Co.) is similar in principle to the Hirst spore trap. It has a built-in vacuum pump, and samples for a week without attention on an adhesive-coated transparent tape on a clock-driven drum behind the entrance orifice (Fig. 6-17). After exposure, the tape is cut into daily segments which are mounted on microscope slides for examination. Other advantages and disadvantages are similar to those of the Hirst spore trap. It is also available commercially. Other samplers of similar design to those described above were reported by Panzer et al. (1957), Voisey and Bassett (1961), and Schenck (1964), but none is known to be commercially available.

Cascade impactor

The cascade impactor was first described by May (1945), but several models by different manufacturers are now available. It consists of a series of slit-shaped orifices in series, each smaller than the preceding. Behind each orifice is a slide

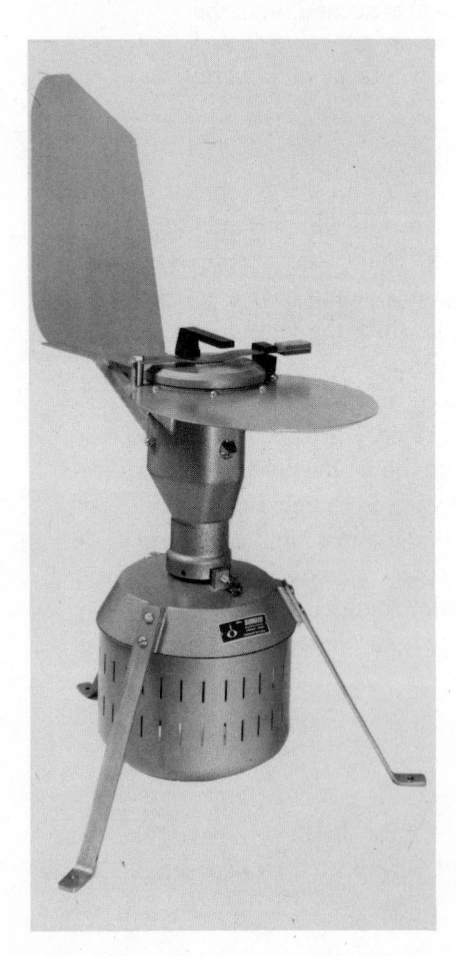

FIGURE 6-17 The Burkard spore trap. Particles are sucked into the orifice beneath the rain shield and impacted on adhesive-coated tape wrapped on a drum which rotates behind the orifice over a weekly period. The lower housing contains a motor and vacuum pump. The tail keeps the orifice facing the wind.

or other deposition surface. Air is drawn through the device at a constant rate giving progressively higher speeds through each orifice. Large particles impact in the first stage while each succeeding stage captures a progressively smaller size class. Size discrimination is not absolute since a considerable overlap exists between stages. The original model had four stages and sampled 17.5 liters/minute giving speeds of 2.2, 10.2, 20.4, and 34 mps through the slits. Models are now available with moving slides for time discrimination, with up to six stages and with a filter as the final stage for capture of the smallest particles. The sampler requires an external vacuum pump. Although the sampler efficiently collects particles which enter the outer orifice, a representative sample of large particles can only be expected to enter if the slit is facing a wind approaching at the entrance speed. Cascade impactors are available commercially at moderate prices but are not suggested for pollen sampling unless size discrimination is desired. Somewhat similar samplers are available which deposit the particles on dishes of nutrient media for culture of spores and bacteria. The best known is the Anderson sampler (Anderson 1958) which is commercially available.

Other nonfilter suction samplers

Another class of suction samplers has been used in which a small intake orifice is located on top of the housing with an impaction surface below. These are characterized by very poor entrance efficiency, although some contain other excellent features. The Marx impinger (Marx et al. 1959), for instance, impacts the sample on a large circular plate which rotates horizontally under a 1 × 10 mm orifice. The plate revolves once in 24 hours and is marked off in one-hour segments giving good time discrimination. The sampling rate is 10 liters/min.

A sampler somewhat similar to the Hirst and Burkard spore traps, but with the entrance orifice on the top, has been described (Pady 1959). It contains a slide which moves in discrete steps so that 24 separate bands are deposited on a single slide at the rate of one per hour. A newer version of this sampler was described by Kramer and Pady (1966) and is commercially available as the Kramer-Collins spore sampler. It samples 22.7 liters/min through a 14 × 0.75 mm orifice.

High-volume filter samplers

High-volume samplers consist of a vacuum cleaner-type motor and blower in a housing with a filter holder on the intake end. Holders are available for four-inch-diameter circular filters and 6 × 9 and 8 × 10-inch rectangular filters. The first such sampler was described by Silverman and Williams (1946). Similar samplers are now produced by several manufacturers and are commonly used in air pollution studies. Flow rates range up to 70 ft^3/min but vary with the type of filter used. Most filters are fibrous types in which collected particles imbed, making them unsuitable for pollen collection. However, glass fiber filters may be used which give a flow rate of about 20 ft^3/min and which retain pollen on the surface. This low flow rate may cause overheating if sampling is continued longer than an hour. Since the high volume sampled is gained by use of the large filter area, linear flow through the filter is quite low. Thus, the sampler is subject to the same entrance problems for large particles as other suction devices. Its other chief disadvantage for pollen counting is the difficulty of counting particles on, or removing them from, the large filters.

Other filter samplers

Molecular membrane filters, which are available from at least two sources, and Nuclepore filters retain particles on the surface where they may readily be examined under the microscope. They are available in various diameters, but the one-inch (25 mm) and the 47 mm sizes are most commonly used. They also come with various pore diameters, but those with the larger pore sizes (5 to 10 microns) are best for pollen sampling, since they give the highest flow rates. These filters may be mounted in any of a variety of filter holders available commercially. One such holder, assembled and disassembled, is shown in figure 6-18. Suction may be provided by any vacuum pump having a nonpulsating flow and adequate air-moving capacity. Since these filters have a comparatively high resistance and develop a large pressure drop, a pump with too much capacity must be provided with an alternate intake to avoid overheating the pump or rupture of the filter. Since filter samplers can seldom be operated even approximately isokinetically, they are subject to the entrance problems discussed in the last Chapter and may have a variable entrance efficiency for pollen-sized particles.

Advantages of filter samplers are:

1. The samples are easy to handle, store, and examine under the microscope.
2. Filter holders and vacuum pumps are readily available commercially at moderate cost.

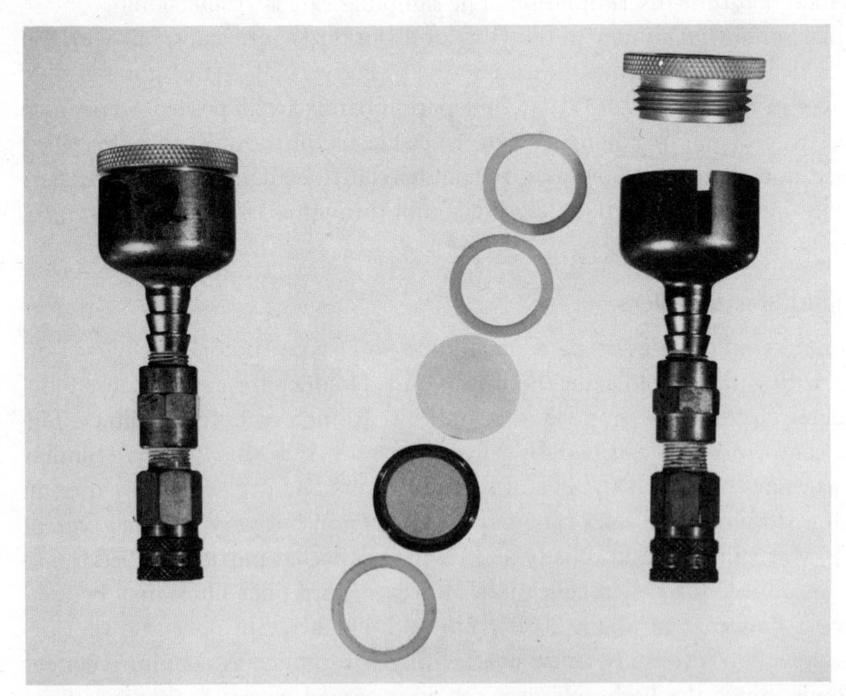

FIGURE 6-18 An assembled and disassembled filter holder. The internal components from bottom to top are: (1) teflon washers, (2) fine screen support, (3) filter, and (4) two teflon washers.

Disadvantages are:

1. The entrance efficiency varies with wind speed, flow rate, particle size and density, and orientation of the entrance with wind direction.
2. Electric power is required.

References

Anderson, A.A. 1958. New sampler for the collection, sizing and enumeration of viable airborne particles. J. Bacteriol. 76:471–484.

Cole, A.L., and A.W. Stohrer. 1966. Sequential roto-disk pollen sampler. U.S. Public Health Serv. Rep. 81:577–578.

Dingle, A.N. 1957. Hay fever pollen counts and some weather effects. Bull. Amer. Meteorol. Soc. 38:465–469.

Durham, O.C. 1946. The volumetric incidence of airborne allergens. IV. A proposed standard method of gravity sampling, counting and volumetric interpolation of results. J. Allergy 17:79–86.

Gregory, P.H. 1973. The microbiology of the atmosphere. 2nd rev. ed. John Wiley & Sons, New York. *See* Chapter 9, Air sampling technique.

Harrington, J.B., G.C. Gill, and B.R. Warr. 1959. High-efficiency pollen samplers for use in clinical allergy. J. Allergy 30:357–375.

Hayes, J.V. 1969. Comparison of the rotoslide and Durham samplers in a survey of airborne pollen. Ann. Allergy 27:575–584.

Henderson, J.J., and W.W. Stalker. 1966. Pollen and spores: comparison of sampling and counting methods. U.S. Dep. Health, Educ. and Welf., Public Health Serv., Div. Air Pollut., Cincinnati, Ohio.

Hirst, J.M. 1952. An automatic volumetric spore trap. Ann. Appl. Biol. 39:257–265.

Kramer, C.L., and S.M. Pady. 1966. A new 24-hour spore sampler. Phytopathology 56:517–520.

Lewis, D.M., and E.C. Ogden. 1965. Trapping methods for modern pollen rain studies. Pages 613–626 *in* B. Kummel and D. Raup, eds. Handbook of paleontological techniques. W.H. Freeman & Co., San Francisco.

Magill, P.L., E.D. Lumpkins, and J.S. Arveson. 1968. A system for appraising airborne populations of pollens and spores. Amer. Ind. Hyg. Ass. J. 29:293–298.

Marx, H.P., J. Spiegelman, and G.I. Blumstein. 1959. An improved volumetric impinger for pollen counting. J. Allergy 30:83–89.

May, K.R. 1945. The cascade impactor: an instrument for sampling coarse aerosols. J. Sci. Instrum. 22:187–195.

Metronics Associates. 1967. Product Bulletin No. 17–67. Metronics Associates, Palo Alto, Calif.

Ogden, E.C., and G.S. Raynor. 1960. Field evaluation of ragweed pollen samplers. J. Allergy 31:307–316.

Ogden, E.C., and G.S. Raynor. 1967. A new sampler for airborne pollen: the rotoslide. J. Allergy 40:1–11.

Pady, S.M. 1959. A continuous spore sampler. Phytopathology 49:757–760.

Panzer, J.D., E.C. Tullis, and E.P. Van Arsdel. 1957. A simple 24-hour slide spore collector. Phytopathology 47:512–514.

Perkins, W.A. 1957. The rotorod sampler. 2nd semiannual report, CML 186. Aerosol Laboratory, Stanford Univ., Stanford, Calif.

Raynor, G.S. 1972. An isokinetic sampler for use on light aircraft. Atmos. Environ. 6:191–196.

Raynor, G.S., and E.C. Ogden. 1970. The swing-shield: an improved shielding device for the intermittent rotoslide sampler. J. Allergy 45:329–332.

Raynor, G.S., J.V. Hayes, and E.C. Ogden. 1970. Experimental data on ragweed pollen dispersion and deposition from point and area sources. BNL [Brookhaven National Laboratory, Upton, New York] 50224 (T–564).

Rosinski, J. 1957. Some studies on the evaluation of gummed-paper collectors used in determining radioactive fallout. Amer. Geophys. Union Trans. 38:857–863.

Schenck, N.C. 1964. A portable, inexpensive, and continuously sampling spore trap. Phytopathology 54:613–614.

Silverman, L., and C.R. Williams. 1946. An apparatus for rapid sampling of large air volumes for industrial air analyses. J. Ind. Hyg. Toxicol. 28:21–25.

Solomon, W.R. 1967. Techniques of air sampling. Pages 326–339 *in* J.M. Sheldon, R.G. Lovell, and K.P. Mathews. A manual of clinical allergy. 2nd ed. W.B. Saunders Co, Philadelphia.

Solomon, W.R., A.W. Stohrer, and J.A. Gilliam. 1968. The "fly-shield" rotobar: a simplified impaction sampler with motion-regulated shielding. J. Allergy 41:290–296.

Voisey, P.W., and I.J. Bassett. 1961. A new continuous pollen sampler. Can. J. Plant Sci. 41:849–853.

Webster, F.X. 1963. Collection efficiency of the rotorod FP sampler. Tech. Rept. 98. Metronics Associates, Inc., Palo Alto, Calif. 50 p.

Webster, F.X. 1968. The flourescent particle atmospheric tracer technique. Pages 48–80 *in* J.Y. Wang, ed. Proceedings of the conference on air pollution in California. San Jose State College. San Jose, Calif.

7

LOCATION OF SAMPLERS

SELECTION OF SITES

Equally important as choosing the proper sampler is the selection of a site (or several sites) that will acceptably fulfill the following criteria:

1. Representative of the area.
2. Free from contamination.
3. Accessible.
4. Free from public liability.
5. Free from vandalism and thievery.
6. Sampling equipment acceptable in the area.
7. A source of electric power, if needed.

The concentration of pollen at the site must be indicative of the area it represents; whether the immediate area, a few hundred feet away, or several miles distant. Obviously, if an average concentration at the immediate location will suffice, the sampler may be anywhere: outdoors, in a building, at any altitude, and without regard to buildings that might affect the flow of air and thereby the concentrations of pollen. If the data from the samples are to indicate the concentrations beyond the site, great care must be taken to assure that the site is representative of the wider area. At ground level (or nose level of pedestrians on the streets of a large city), pollen concentrations may vary greatly at locations a hundred feet, or less, away. The effect of air flow with regard to streets and tall building complexes coupled with local pollen sources may result in patterns of pollen concentrations that are far from uniform. Ideally, but generally not practical, a large number of samplers at as many sites should be employed. These sites would be chosen to represent the diversity of situations. An alternative to many samplers at ground level is to choose a site above most of the buildings where the pollen concentrations usually are more uniform over a citywide area. This may be useful for comparing different cities and different days, although it may not indicate closely the concentrations at lower levels.

Contamination from particles other than the pollen being sampled is sometimes necessary to avoid. A site near a dusty road or downwind of a smoky chimney may yield samples that are difficult to analyze. A site comes to mind that was on a roof parapet where the janitress daily shook her dust mop. Not until counting of pollen began was this realized. In some situations, fungus spores are so abundant that the sampling periods may need to be shortened. Even contamination by

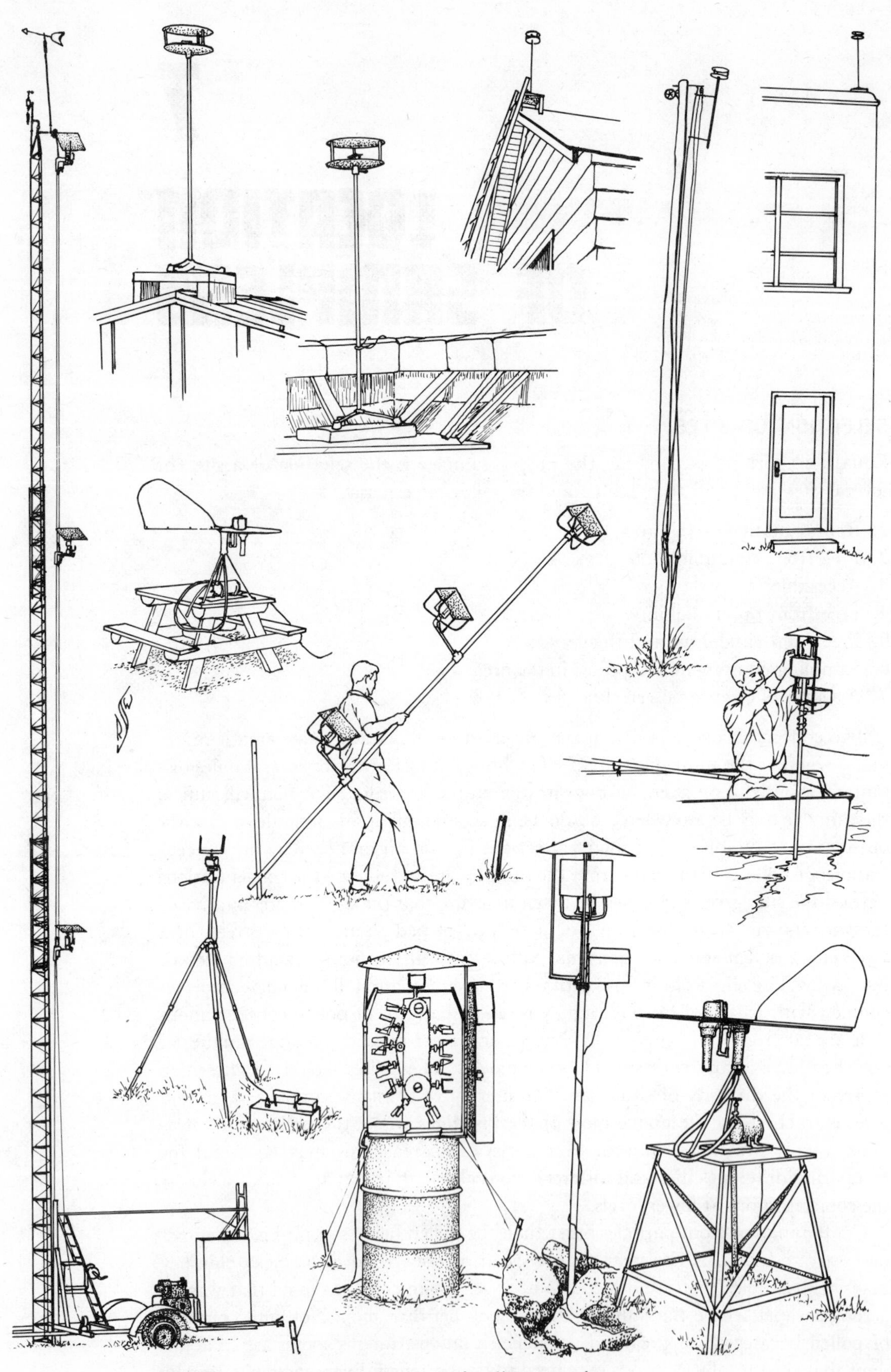

FIGURE 7-1 Some installations of pollen samplers.

pollen of the type being sampled can occur. Proximity to a source may need to be avoided unless it is desired to determine the contributions from that source. A sampler in a patch of ragweed is hardly representative of the general area.

The sampler should be safely accessible to the operator. Daily attendance engenders carelessness. Unsafe ladders and awkward positions may be hazardous, even for the athletic. Samplers that are easily accessible to the operator may be accessible also to unauthorized persons. Precautions should be taken to avoid injury (and possible lawsuits) to curious, careless persons, especially children.

Vandalism and thievery are expensive and, often of more importance, may cause a loss of irreplacable data. Samplers and accessory equipment to be unattended for periods of time may require special protection.

Sites should be avoided where the sampling equipment is annoying because of appearance, noise of the motors, etc. These criteria may seem elementary, but at many observed sampling sites it is obvious that some of them have been ignored.

Most reliable pollen samplers require electric power. If not available at the immediate site, this might be supplied via extension cord from a few hundred feet distant. In special situations, a portable power source can be used, such as a gasoline-powered generator or DC battery and inverter.

TYPES OF INSTALLATION

Small samplers, such as the Durham sampler and the rotoslide, can be mounted on rods or pipes attached to a base of ¾-in. outdoor plywood, 18 to 24 inches square. Commercial Durham samplers are supplied with a three-legged base which may be screwed or bolted to the wood. If the heavy base is not used (as with a homemade sampler), a floor flange to fit the upright rod is suitable. The rotoslide should be mounted on a ¾-in conduit pipe, which is threaded at both ends. For this, use a ¾-in. threaded floor flange attached to the center of the platform. Sand bags or rocks may be used to prevent the stand from tipping in a heavy wind. A cross made from 2 × 4 lumber is better for some situations, especially if weights are not at hand. This should be three or four feet across. For some situations, as roof parapet or slant roof, other base attachments may be needed. If mounting on a narrow parapet, it may be wise to drill holes (as with a star drill) and fasten with expansion bolts and lag screws.

For mounting a small sampler more than two or three meters above the base, two pipes pivoted together can be used to allow the sampler to be brought down for changing the samples. This may be satisfactory to five or six meters, but beyond this height wind action may make more rugged masts necessary. Those taller than 15 meters may require guy wires. A rugged triangular frame mast can be fitted with rope-and-pulley elevator platforms. More elaborate towers include platforms raised and lowered on greased tracks by means of electrically powered winches. Mobile trailer-mounted telescoping towers are commercially available.

Larger and heavier samplers, like the Hirst spore trap, can be bolted to a wooden base and perched on a barrel, box, or table and guyed in three directions. A more satisfactory, but more expensive, stand can be constructed of aluminum. The Hirst trap legs are bolted to the top. The flaring legs of the stand eliminate the need for guy wires.

Sampling from aircraft poses several problems with regard to installation and is not discussed here.

8

SAMPLING TECHNIQUES

Many devices are currently being used to sample airborne pollen. These are discussed in Chapter 6. An attempt is made here to outline a few of the techniques employed by the operator or technician for some of these sampling instruments. Variations in these techniques, due to individual preferences, are acceptable if the quality and representativeness of the sample are not adversely affected. It is the task of the technician to deliver acceptable data and to keep adequate records to aid in the interpretation of these data. Before using the procedures described below, the technician should become familiar with the physical operation of the instrument in use by reference to Chapter 6 and to information supplied by the manufacturer. Some topics and procedures common to more than one sampler are discussed first. These are followed by detailed directions for use of selected samplers.

ADHESIVES

Many pollen samplers require an adhesive to retain the collected particles on the sampling surface. Silicone stopcock grease or silicone oil (Dow Corning Corp.) are recommended for routine use but others, in use by some investigators, include Lubriseal stopcock grease (A.H. Thomas Co.), diluted rubber cement, a pressure sensitive adhesive (Dow Corning 269), petroleum jelly, and glycerin jelly. The latter two are not satisfactory for outdoor sampling because they are extremely sensitive to changes in humidity and temperature, causing variation in their retention efficiency. Dow Corning 269 is satisfactory for samples to be examined unstained or to be stained within a few hours after the samples are taken. Much of the pollen fails to accept a stain after remaining on DC 269 more than a day or two. To apply this adhesive, dilute it with an equal volume of xylene and dip the slide edge to a depth of a few millimeters. Diluted rubber cement is satisfactory for short-period samples, but its retention efficiency appears to decrease before the end of a 24-hour day. Lubriseal stopcock grease has been reported to retain pollen as well as silicone grease and to be more compatible with aqueous mounting media (Solomon et al. 1968). Its melting point (40° or 50°C, depending on the kind chosen) is far below that of DC silicone grease (200°C), so it may not be suitable in hot locations.

Silicone grease is more widely used than any other adhesive for pollen sampling. Its adhesive properties do not change in the ranges of temperature and humidity encountered in nearly all situations. However, it must be carefully applied when using a rotating impactor type of sampler. If the layer is too thick, the pollen grains will tend to be buried and will not stain. If too thin, retention is poor and variable, resulting in serious errors. Ideally, the layer should be 10 microns thick or, at least, between 5 and 15 microns. Uniform application of the grease is not easy to attain.

Silicone oil (DC 200 fluid) may be obtained in a wide range of viscosities. Its great advantage is that the slide edges may be dipped in it, thus yielding uniform results by different technicians. A density of 60,000 centistokes is satisfactory. It is miscible with silicone grease, so a greater density may be obtained when desired. The dipped layer is thick (approximately 1,000 microns), but this almost immediately drops to about 10 microns upon rotation of the slides. Retention of particles in the size range of pollen is essentially the same as with grease. Uniformity of catch is generally better. There is some tendency for the oil to run down the sides of the slides but, if the recommended technique is followed, this is not serious. A small amount of silicone oil will handle a great many samples.

The adhesives mentioned above retain the samples adequately in dry weather but often lose some or all of the pollen when rotated rapidly in rain or fog. A pressure-sensitive adhesive used on AEC fallout paper (Simons 910) has excellent retention in wet air, but its efficiency of catch on rotating slides is low and variable.

STAINING

The techniques described here employ basic fuchsin dye to stain the collected pollen grains. This dye is deposited on the outer wall enhancing the surface characteristics necessary for identification. Basic fuchsin generally does not stain fungus spores nor most of the other debris usually present on samples exposed outdoors. The stain is carried in Calberla's solution or in glycerin jelly.

1. Modified Calberla's solution: 5 ml glycerin, 10 ml 95% ethyl alcohol, 15 ml distilled water, 2 or 3 drops saturated aqueous basic fuchsin, and 2 or 3 drops of melted glycerin jelly.
2. Glycerin jelly: See formula in Chapter 4. Glycerin jelly may also be purchased from scientific supply houses. If glycerin jelly is to be stored for several months, it should be refrigerated. Add one drop of saturated aqueous basic fuchsin for each 15–20 ml of glycerin jelly. More or less stain may be used if desired.

Other dyes used for staining pollen grains are discussed by Wodehouse (1935), Venning (1954), and Brown (1960). Discussion of the characteristics and uses of biological stains are given by Conn (1961) and Gurr (1962).

LABELING AND DIARY NOTES

In any sampling program, it is necessary to label samples and data sheets adequately. Samples should be labeled with the date and often a coded reference to the sampling site and individual instrument. It may be desirable to keep a diary which would include such items as those listed below.

1. Descriptions of the samplers.
2. Descriptions of the sampling sites.
3. Codes used in labeling the samples.
4. Dates and times of sampling periods.
5. Observations concerning the operational condition of the instruments.
6. Descriptions of mechanical failures or power interruptions with interpretations of their effect on the accuracy of the samples.
7. Weather observations, especially precipitation.
8. Any observations which may have an effect on the accuracy of the samples.

Comments which may seem unimportant at the time often prove to be significant later when the data are being analyzed. Also, included in the diary or a separate journal may be descriptions of the sample preparation and counting procedures. The technician making the counts should be alerted to make note of any unusual appearance of the collected sample such as uneven distribution of particulates or an uncommonly clean or dirty portion of the sample.

MICROSCOPY

The use and care of the microscope are described in several available publications. Some books on the microscope, such as Möllring (1968) and Hartley (1964), and chapters in books on botanical microtechnique, such as Sass (1958) and Gray (1964), are helpful. The following brief remarks refer mostly to routine examination of pollen samples but with additional remarks pertaining especially to fungus spores.

A compound microscope is essential. It should be binocular. It should have a condenser, a calibrated mechanical stage, 10× oculars, 10× objective, and high-dry objective (usually about 45×). A whipple micrometer disk in one of the oculars is useful. This divides the large viewing field into smaller areas and often aids the counting process. If samples on slide edges are to be examined, make sure that the microscope chosen has sufficient clearance to handle them. For routine identification, an inexpensive substage lamp is adequate. However, the clarity of image needed for the detection of some characters will require one of the better illumination systems, preferably one built into the microscope stand.

The magnification used depends on the frequency, size, and other characteristics of the pollens to be identified and counted. Easily recognized pollens can, with practice, be identified at magnifications from 100 to 200 diameters. Those grains that are not adequately seen at the lower magnifications should be examined at 400× or higher. It is common practice to scan at 100× with frequent shifts to 400× for better views of some grains. For some samples, still higher magnifications (obtained with an oil-immersion objective) may be desirable.

The depth of field (depth of focus) at 100× or more is much less than the diameter of even the smallest pollen grains. To properly examine the curved surface at all levels, one should continually vary the focus, using the fine-adjustment knob. This also helps to bring into view pollen grains at different levels in the sample.

More critical microscopy may be needed for the identification of some fungus spores because of their smaller size. Routine counting of familiar spores may be done with the high-dry objective (40-60×), but identification of unknowns should be with an oil-immersion objective and optimum condenser illumination. The condenser must be focused for best results, and standard condensers will not come to a sharp focus through a slide on edge; therefore, limited resolution is obtained for collections made on slide edges.

For oil-immersion lenses, the working distance (the distance between the bottom of the objective and the object being viewed) is very small, and it is often impossible to focus on objects in a thick adhesive. Mounts should be made thin if oil-immersion observation is anticipated.

Specialized stains are usually not necessary for identification of fungi. They may hide subtle pigmentation and make identification more difficult. Fungus spores mounted in Calberla's solution containing basic fuchsin normally do not accept the stain and are suitable for identification. It is helpful but not necessary to have a phase contrast microscope for accentuating minute features of unpigmented spores; however, subtle pigmentation may be masked.

A technique or "trick" which may be useful in the microscopy of fungus spores is oblique illumination. Septa, ridges, flagella, or other features which form a straight line may be accentuated by illumination with a slit of light parallel to the object. This may be accomplished by placing a piece of dark paper with a slit cut in it between the light source and the substage condenser. A similar effect can be produced by placing the edge of a piece of paper or the edge of your hand in the same location. The effect of this is a distortion of the image reducing the visibility of features perpendicular or oblique to the slit but enhancing the visibility of those parallel to it. This is particularly useful for distinguishing thin septa in spores.

DURHAM SAMPLER

The Durham sampler, also described as the gravity slide sampler (Durham 1946), has been the standard sampler recommended by the American Academy of Allergy for many years.

List of supplies

Durham sampler (Fig. 6-1).
Microscope.
Glass slides, 25 × 75 mm, preferably with frosted ends or other label which may be marked in pencil.
Cover glasses, 22 × 22 mm. Thickness No. 1 is a good choice.
Hand tally counter.
Slide boxes.
Silicone stopcock grease.
Fine-pointed forceps and dissecting needle.
Medicine dropper (dropping pipet) and bottle.
Staining solution.

Preparing the slides

1. Label the slides on the frosted end, showing location and date of the sample.
2. Apply silicone grease to the flat surface of the slide. Spread the grease over the

entire surface, excluding the label, with the forefinger. Then rub the grease to a smooth, thin layer with the heel of the hand.
3. Store the slides, until used, in a standard slide box.

Taking the sample

1. Transfer the slide to a separate box for transport to the sampling station. This is to prevent contamination of the other slides.
2. Place the slide on the slide platform, slipping one end under the clip.
3. After exposure pull the slide out from under the clip, being careful not to touch the collecting surface.
4. The slides (samples) may be stored indefinitely before they are counted.

Preparation for counting

1. Remove any large pieces of debris from the sampling surface with fine-pointed forceps.
2. Using a medicine dropper, place a few drops of Calberla's solution directly on the slide.
3. Apply the cover glass. Place one edge of the cover glass on the slide and lower the glass with a lever action, supporting the cover glass on a straight dissecting needle as it is lowered, allowing as much air as possible to escape from underneath.
4. The pollen grains will be stained in 3–5 minutes.

If it is desirable to retain the samples for rechecking, they should be mounted in glycerin jelly to which the stain has been added. The jelly may be stored at room temperature or in a refrigerator. For use, it is warmed in hot water until it flows from a dropper (Fig. 4-2). Such samples require several hours for the stain to take effect and the jelly to harden. When cool, they may be kept for long periods without drying, and the cover slip would be held more firmly to the slides.

Counting

1. Place the slide on the microscope stage and sweep the stained portion from one edge of the cover glass to the other, using 100× magnification. Shifts to a higher power may be made for identification of some grains.
2. Tally the number of grains encountered.
3. Several sweeps distributed across the slide are required to cover a sufficient area. For routine counting, approximately two square centimeters are usually sufficient. The area of each sweep is determined by multiplying the width of the field of view by the length of the sweep.
4. Divide the count by the number of cm^2 actually examined to obtain the number of grains on one cm^2 of sampling surface.

ROTOSLIDE

The methods suggested here have been developed during several years of sampling with the rotoslide and with other samplers in which the sample is taken on the edge of the slide. Unless otherwise indicated, the following suggestions refer to the intermittent swing-shield rotoslide. Modifications to suit individual preferences will occur to many.

List of supplies

Rotoslide sampler (Fig. 6-13).
Slide positioner (Figs. 8-4 and 8-5).
Microscope. Be sure it has sufficient clearance to handle slide edges.
Glass slides, 25 × 75 mm, preferably with frosted ends or other label which may be marked in pencil. These must be without beveled or oblique edges and should be approximately one mm thick.
Cover glasses, 22 × 50 mm. Thickness No. 1 is a good choice.
Tubular micrometer caliper, calibrated in metric system.
Hand tally.
Slide boxes.
Material for modifying slide boxes.
Silicone stopcock grease or silicone oil (60,000 centistokes).
Fine-pointed forceps.
Medicine dropper (dropping pipet) and bottle.
Staining solution.

Modification of slide box

Choose a wooden slide box with a deep cover and modify it to prevent the sampling edge of the slide from hitting against the cover when the box is closed. Strips of closed-cell foam polyurethane, or rubber weather stripping, or rubber tubing may be fastened onto the inside of the cover so they touch the corners of the slides when the box is closed (Fig. 8-1). Only the central 50 mm of the length of each slide edge will be examined, so a few millimeters at each end of the sampling surface may be sacrificed. We prefer the polyurethane, but have found that in contact with silicone oil, and to some degree with silicone grease, it eventually loses its resilience and needs to be replaced.

Another method that leaves the slide edges untouched, even at the ends, is illustrated in figure 8-3. The wooden dowel, that is glued to the cover, presses against the hollow rubber tubing, that is glued to the center strip between the slides, expanding the tubing horizontally and pushing the slides firmly to the opposite walls. This prevents the slides from hitting the cover when it is closed, but allows free insertion and removal when the cover is open.

Preparing the slides

1. Check the slide edges to be used for the sampling surface with a magnifying glass to be sure they are cut at right angles to the flat surface (Fig. 8-1). If one edge is satisfactory, the other edge need not be examined. Indicate the chosen edge on the frosted end. Do not use those which are oblique, because the field of view under the microscope will be partially out-of-focus, and the liquid mounting solution may dry out too quickly.
2. Measure the thickness of the slides with the micrometer caliper (Fig. 8-1).

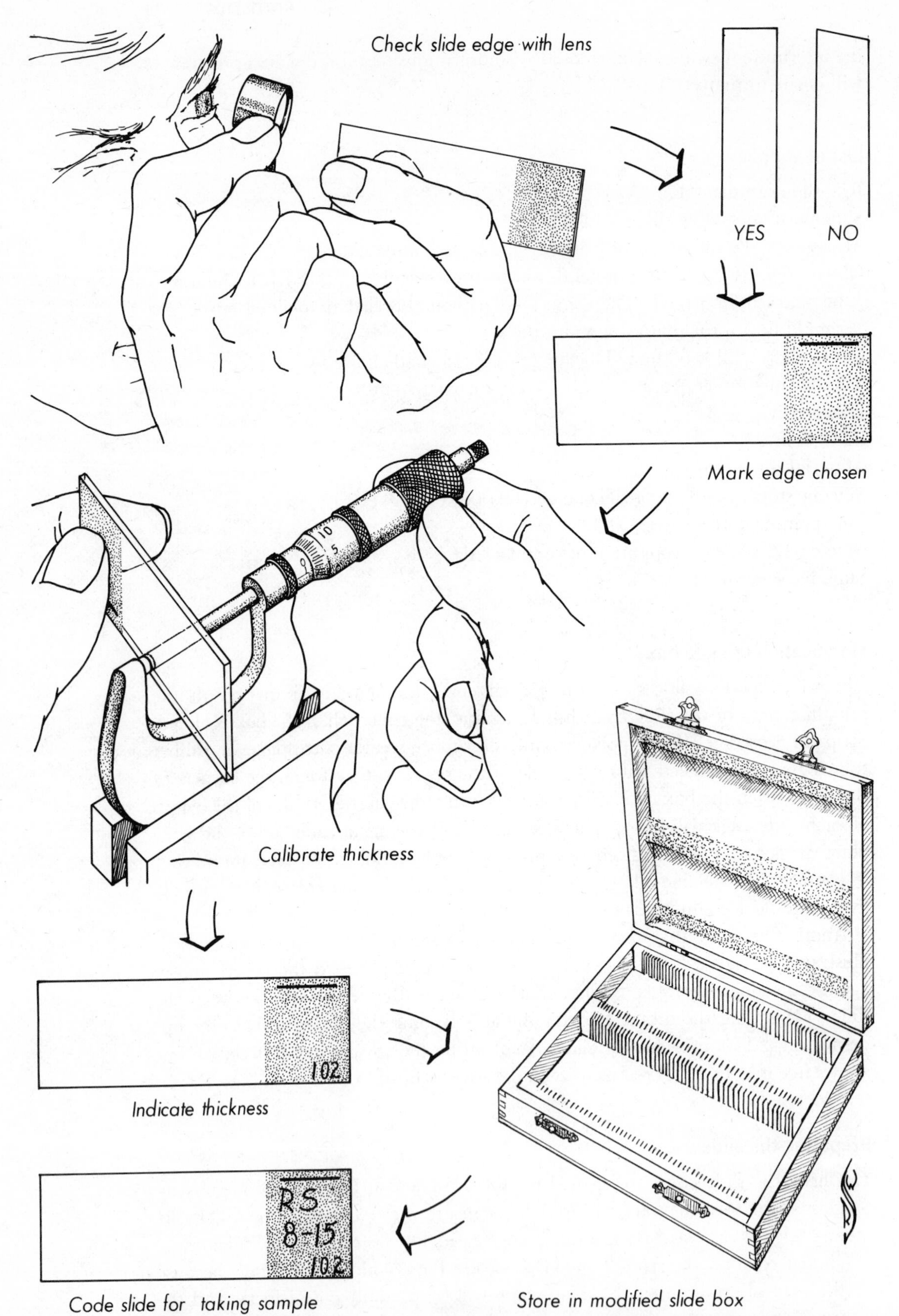

FIGURE 8-1 Procedure for preparing microscope slides for slide-edge samples.

Only the edge to receive the sample needs to be measured. Discard slides less than 0.90 mm and more than 1.10 mm. Record the thickness to the nearest hundredth of a millimeter on the frosted end. Microscope slides often vary in thickness beyond the tolerances listed by the manufacturer. If slides are obtained which do not vary much in thickness, this operation may be omitted.

3. Label the slides on the frosted end, showing location and date of the sample (Fig. 8-1). Two slides comprise each sample.
4. Place the slides, sampling edge up, in the modified slide box (Figs. 8-1 and 8-3).
5. Apply adhesive.

   Silicone grease method (Fig. 8-2). Several slides may be greased at one time without removing them from the modified slide box. Put a small amount of silicone grease directly from the tube or jar, or from some squeezed onto a piece of paper, on the forefinger. Spread the grease smooth by lightly rubbing the forefinger against the thumb. Apply the grease to the slide edges by rubbing the forefinger lightly across the slides as they stand in the box. Make sure the entire length of each slide is contacted. Then with a clean finger rub the edges lightly again to smooth the grease. This should result in a fairly uniform layer of grease from 5 to 15 microns thick. Any overflow of grease may be removed with a small spatula.

   Silicone oil method (Fig. 8-3). Pour some of the oil into a flat glass dish, such as a petri plate, to give a depth of 5 mm or more. Dip each slide individually so the sampling edge touches the oil evenly. The slide may penetrate for a few millimeters. Remove the slide and hold, for a second or so, in such a way that the dipped edge is semihorizontal allowing excess oil to run to the lower end, where it may be removed (if necessary) with absorbent paper or the rim of the petri plate. Turn over the slide and place it into a slide box so the dipped edge is up and horizontal. This process takes but little time and is faster than proper application of grease. After a few minutes, remove large excesses of oil (if any) from the sides of the slides with a suitable tool such as a scalpel or thin cardboard, wiping the tool with absorbent paper such as facial tissue. Some oil may run down the slides but that is not overly objectionable.
6. The greased or oiled, labeled slides are stored in the modified slide box until used.

Taking the sample

1. Transfer some slides to a separate box for transport to the sampling station. This is to prevent contamination of the other slides.
2. Set the switch to the *off* position.
3. Remove the slide from the box and hold it with the thumb and forefinger, taking care not to touch the central 50 mm of the sampling surface (Fig. 8-2).
4. With the other hand, grasp the slide holder and swing a shield to permit insertion of the slide.
5. Slip the slide into the holder (frosted end down) and push it back until its whole length is against the back of the holder. The sampling edge should be in a straight vertical position.

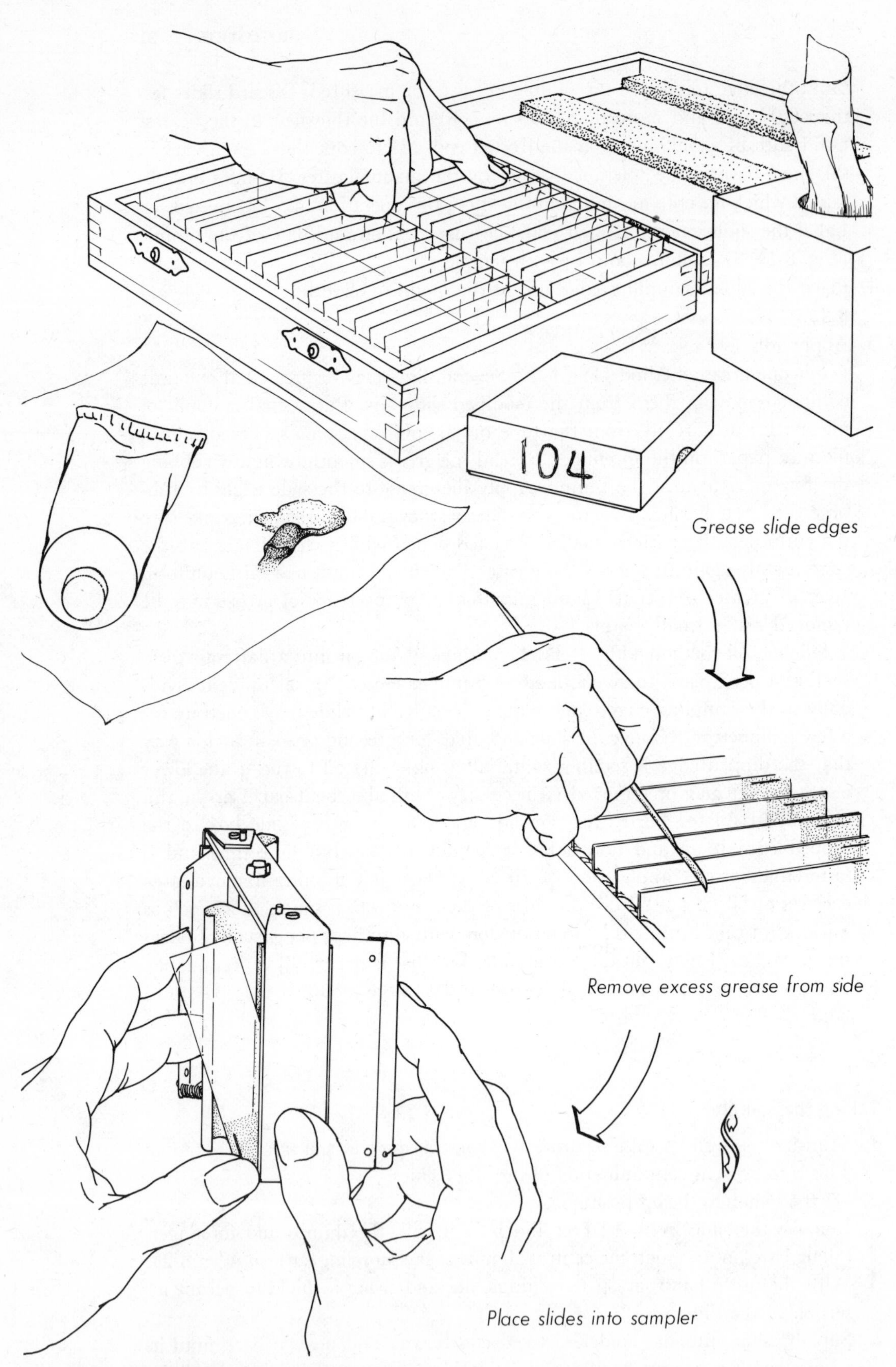

FIGURE 8-2 Applying silicone grease to slide edges and placing slide into roto-slide sampler.

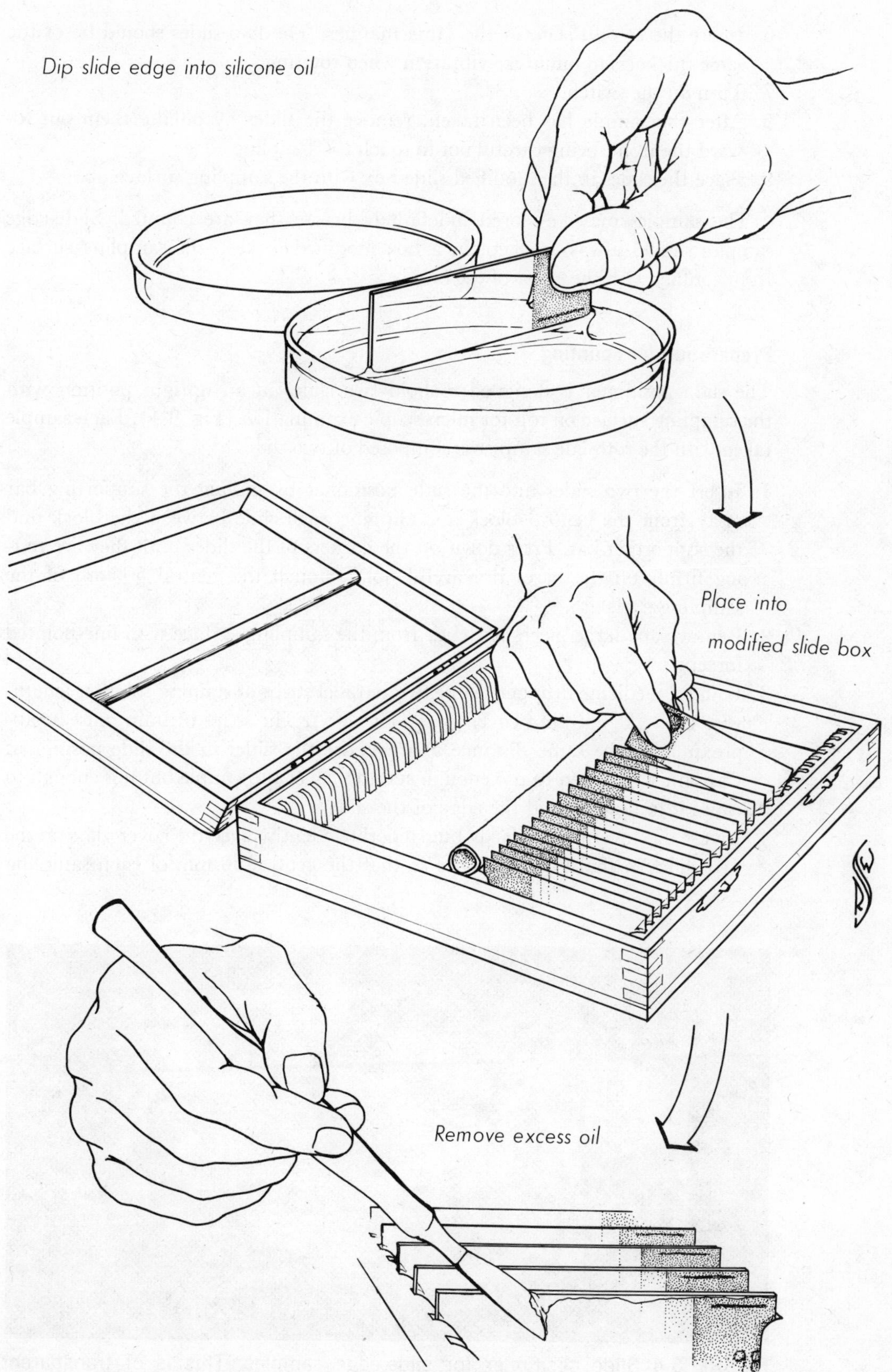

FIGURE 8-3 Applying silicone oil to slide edges.

6. Insert the second slide in the same manner. The two slides should be of the same thickness to minimize vibration when rotating.
7. Turn on the switch.
8. After the sample has been taken, remove the slides by pulling them out toward the front, being careful not to touch the sampling edge.
9. Place the slides in the modified slide box with the sampling surface up.

The samples may be stored indefinitely before they are counted. Slide-edge samples should always be kept in a box modified to keep the sampling surface from contact with the cover of the box.

Preparation for counting

The slide positioner is designed to hold two slides in an upright position with the sampling surface on top for microscopic examination (Fig. 8-4). Each sample taken with the rotoslide sampler is composed of two slides.

1. Insert the two slides into the slide positioner by pulling the supporting bar away from the central block and slipping each slide between the block and the supporting bar. Press down on the corners of the slides until they are resting firmly on the base. Be careful not to touch the central 50 mm of the sampling surface.
2. Remove any large pieces of debris from the sampling surface with fine-pointed forceps.
3. Using a medicine dropper, paint two parallel strips of staining solution lengthwise on a 22 × 50 mm cover glass (Fig. 8-5). The strips of stain must be approximately the same distance apart as are the slides in the slide positioner. Care must be taken to use enough stain to cover the sample but not enough to cause it to flow beyond the edge of the slide.
4. Invert the cover glass quickly but smoothly. Gently place the cover glass on the two slides in the slide positioner so that the central 50 mm of each sampling

FIGURE 8-4 Slide positioner for slide-edge samples. This is of transparent plastic to allow light from the microscope lamp to pass through freely.

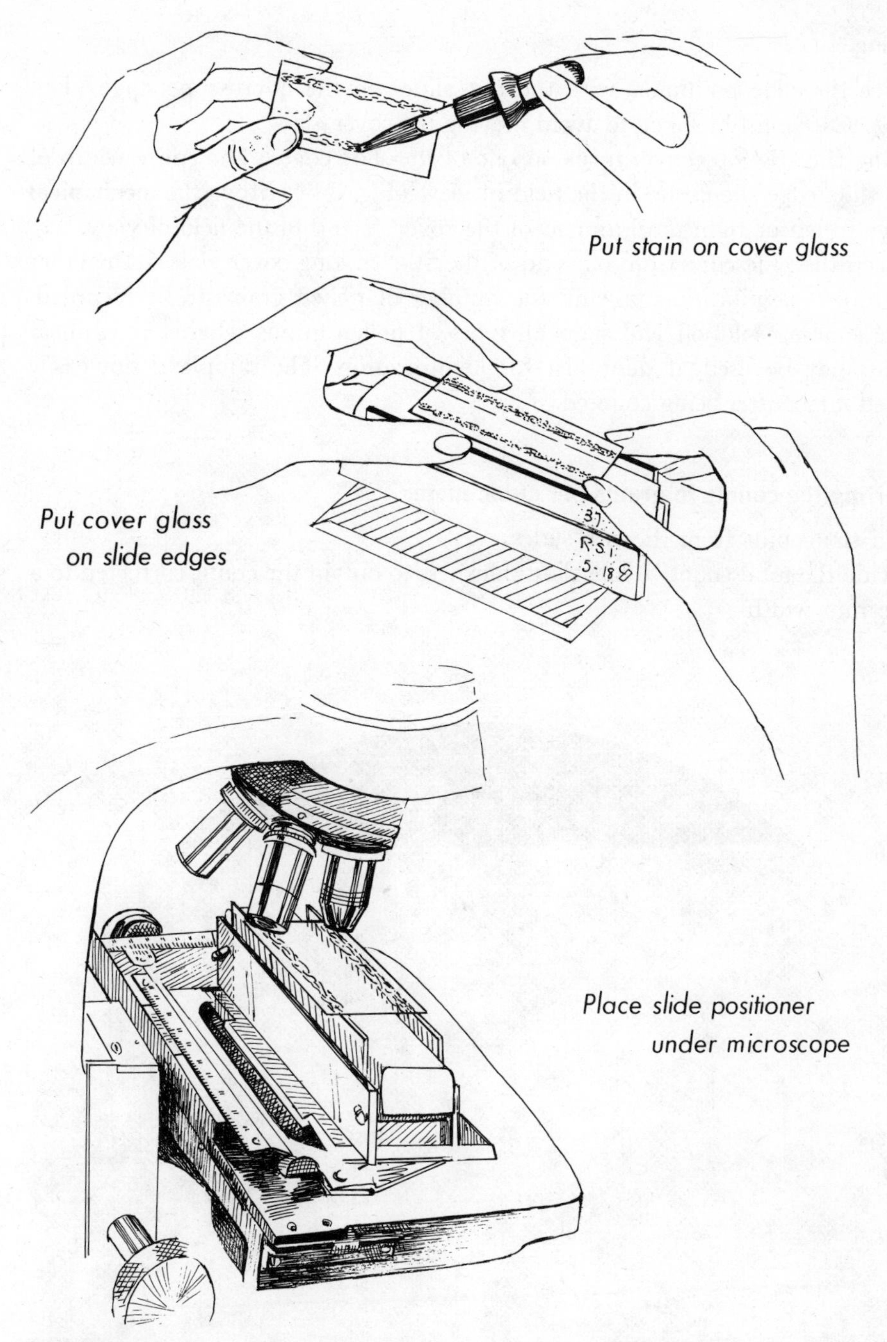

FIGURE 8-5 Preparing slide-edge samples for examination.

surface is covered with the stain (Fig. 8-5). The pollen grains will be stained in 3–5 minutes.

An alternate method of preparing the sample is: Instead of Calberla's solution, use glycerin jelly to which the stain has been added. The jelly may be stored at room temperature or refrigerated but must be warmed in hot water until it will flow from a dropper (Fig. 4-2). When cool, such samples may be kept for long periods without drying, and the cover slip is held more firmly to the slides.

Counting

1. Place the slide positioner with the two slides on the microscope stage (Fig. 8-5). Care must be taken to avoid moving the cover glass.
2. Using the 10× objective, focus on one of the slide edges. The entire width of the slide edge should be in the field of view (Fig. 8-6). Move the mechanical stage to left or right until an end of the cover glass is in the field of view.
3. Sweep the slide edge from one end of the 50-mm-long cover glass to the other at 100× magnification tallying the number of pollen grains to be recorded. The staining solution will stain all types of pollen grains. Higher magnifications may be used, if desirable, for identification. The sample is not easily saved intact after being counted.

Converting the counts to grains per cubic meter

1. Add the counts from the two slides.
2. Divide the total count by the slide thickness to obtain the count corrected to a one mm width.

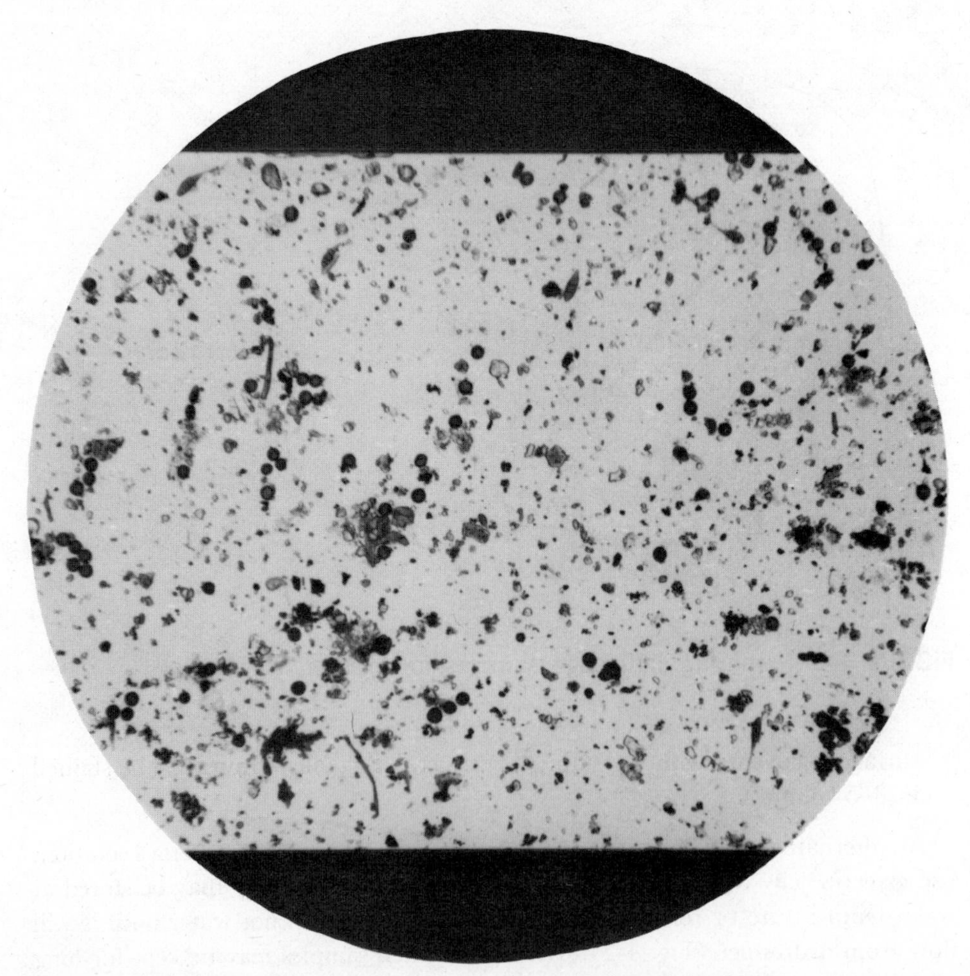

FIGURE 8-6 Slide-edge sample as seen through the microscope.

3. Divide the corrected count by the factor to determine the average number of pollen grains per cubic meter.

The factor is the volume of air that is sampled multiplied by the efficiency of impaction and retention. It may be supplied by the manufacturer of the sampler, or determined by the purchaser, as follows: The area of the sampling surface is the length of the 50 mm cover glass times the 1 mm width of the edge of the slide. The complete sample is the sum of the two slides, an area of 100 mm^2 or 0.0001 m^2. This area multiplied by the linear air travel past the sampling surface is the volume of air sampled. Linear air travel is determined by multiplying: (1) linear travel each revolution which equals the circumference of the circle traversed by each slide holder, (2) number of revolutions per minute, and (3) operating time in minutes. The usual swing-shield intermittent rotoslide sampler has a linear travel each revolution of 0.3585 meters and an rpm of approximately 1620 with the loaded slide holder which is 581 meters per minute. It operates for one minute in each 12 minutes for a total of 120 minutes in a 24-hr. day.

Thus, the volume of air sampled in two hours of activity is 0.0001 (m^2) $\times$ 581 (m/min) $\times$ 120 (min) which is about 7 cubic meters.

The efficiency of the rotoslide sampler varies somewhat with wind speed. Extensive tests in a wind tunnel, with ragweed pollen, indicate the average efficiency in still air to be 70% and at 10 mps (22 mph) to be 50%. In the range of airflows mostly encountered while sampling airborne pollen, the average efficiency is likely to be approximately 64%. Thus, in most situations, the numerical factor for converting the one sq cm surface count to the average number per cubic meter is 64% of 7, being about 4.5. For example, a ragweed pollen count of 100 is divided by this factor which indicates an average of 22 grains per cubic meter. See Fig. 8-7.

Reduced to a simple formula this is: area of sample (m^2) $\times$ distance per revolution (m) $\times$ number of revolutions per minute $\times$ number of minutes $\times$ efficiency (%) = dividing factor to convert the count to the average number per cubic meter of air. This is: $0.0001 \times 0.3585 \times 1620 \times 120 \times 0.64 = 4.5$.

Or, one might sample for ragweed pollen in a room in still air (efficiency 70%) continuously for one hour with a sampler whose rpm is 1570. Derivation of the factor then would be: $0.0001 \times 0.3585 \times 1570 \times 60 \times 0.70 = 2.4$.

When higher magnification is desired for sweeping the sample, the volume of air sampled is determined in the same manner except that the area counted and consequently the volume is reduced. The area width is now the diameter of the field of view.

If a rotoslide sampler is equipped with a revolution counter, derivation of the factor would be: volume per revolution $\times$ number of revolutions $\times$ efficiency. If the sampler holds four slides (Fig. 6-15), instead of two, the volume per revolution is doubled.

Preservation of the sample

Some types of pollen sampling require the preservation of the sample for future reference. When the liquid staining solution is used, removal of the cover glass from the slide edges disturbs the sample and may remove some of the pollen. The loss of pollen by the removed cover glass may vary from 0–20%, with an average

1970 — Saratoga — intermittent roto — ragweed

	COUNTS		SUM	WIDTH	CORRECTED COUNT	AV. PER M^3
Aug. 15	50	45	95	95	100	22
16	63	79	142	105	135	30
17	155	140	295	99	298	66
18	780	695	1475	100	1475	328
19	404	390	794	101	786	170
20	160	135	295	97	304	68

FIGURE 8-7 Sample data sheet for tabulating pollen counts from a rotoslide that operated for a total of two hours with a factor of 4.5.

of approximately 3%. Furthermore, the staining action in this solution is continuous, causing grains to become darker with time and obscuring features of the exine necessary for the identification of different pollen types.

By mounting the sample in glycerin jelly prestained with basic fuchsin, it is possible to remove the sample from the slide edges and store it for future reference on a conventional flat side. The staining action in glycerin jelly is not continuous, although the pollen grains may require several hours to absorb the stain.

The preparation of slide-edge samples for storing has been tested only with samples collected on silicone stopcock grease. If another adhesive is employed, it is recommended that the technique be tested before use.

1. Apply the glycerin jelly to the samples as described in preceding sections of this Chapter. Instead of the 50-mm-long cover glass, two smaller ones (22 × 22 mm) may be used for ease in handling.
2. Chill the mounted slide edges in their slide positioner overnight in a refrigerator.
3. Carefully lift the cover glass, with the sample embedded in the hardened jelly, from the slide edges and place it, sample down, on a clean flat microscope slide. Tap gently a couple of times with a dissecting needle to make the jelly adhere to the flat slide.
4. It is advisable to scan the slide edges from which the sample has been removed to record any pollen grains that may have been left behind.
5. The removed sample may then be stored in conventional slide boxes.

Variable counting areas

For some sampling programs, the scanning of a 50-mm-length of each slide edge may be more time-consuming than is necessary for the required accuracy when the collected amounts of the pollen of interest are substantial. In these cases, selected portions of the total length may be scanned. It is recommended that the portions scanned be distributed along the length of the slide edge. For example, by making reference to a graduated mechanical stage, alternate 5-mm-long portions can be scanned for a total of 20–30 mm. When converting to volumetric measure, the count is divided by the percentage of the 50-mm-length actually examined, or the area of the portion counted is used in the conversion process detailed earlier.

When counting pollens which require greater magnification for identification, the slide edge may be scanned at low power (100×) with shifts to a higher power objective lens for identification of individual grains. If the pollen is closely distributed, the slide may be scanned at the higher power. In this case, only a part of the width of the slide edge is in view so care must be taken to examine a representative portion of the sample. Scanning one sweep in a direction parallel to the long axis of the slide may result in erroneous counts since some pollen grains are moved toward the edge during the staining process. It is then more accurate to scan in adjacent sweeps in a direction perpendicular to the long axis of the slide. The portion scanned is determined by reference to the calibrated mechanical stage or by determining the number of sweeps and the diameter of the field of view of the microscope lenses in use.

Length of sampling periods

The intermittent rotoslide is designed to sample for one minute in every 12 minutes for 24 hours, but the sampler can be used with other sampling schedules. In some situations, it may be desirable to change the slides more frequently. For instance, if rain occurs after some period of clear weather, it might remove pollen already collected. A rotoslide spinning continuously can be used to give a sample up to 2 or 3 hours in length. Several rotoslides operated in tandem allow continuous sampling for longer periods (Fig. 6-14). In each case, the actual spinning time in minutes must be recorded. This time is then substituted for the normal 120 minutes in the conversion of the counts to grains per cubic meter previously described.

HIRST SPORE TRAP

The procedures suggested here are a combination of those described by Hirst (1952) and modifications developed during several years of sampling with the Hirst spore trap.

List of supplies

Hirst spore trap (Fig. 6-16).
Vacuum pump (refer to Chapter 6).
Microscope.
Glass or plastic slides, 25 × 75 mm. If glass slides have frosted ends, the frosted area should not extend more than two centimeters to insure having the sample on clear glass. Plastic slides may be labeled with a felt marker.
Cover glasses, 22 × 50 mm. Thickness No. 1 is suitable.
Hand tally.
Slide boxes.
Silicone stopcock grease.
Fine-pointed forceps, straight dissecting needle, and slender scalpel.

Medicine dropper (dropping pipet) and bottle.
Staining solution.

Preparing the slides

1. Label the slides near one end.
2. Apply silicone grease to the slide. Spread the grease with the forefinger over the entire flat surface of the slide, excluding the label. Rub the grease to a smooth, thin layer with the heel of the hand.
3. Store the slides in a clean slide box until ready for use.

Taking the sample

1. Transfer the slide to a separate box for transport to the sampling station. This is to prevent contamination of the other prepared slides.
2. Turn off the vacuum pump.
3. Insert the bolt at the base of the vane tail to lock it in a stationary position. This may be omitted in quiet air.
4. Mark the previous sample (Fig. 8-8). If glass slides are used, insert a thin stylus through the orifice and scratch the glass or at least make a mark on the adhesive. Some prefer to release a small puff of some recognizable material, such as *Lycopodium* powder (spores) or dye crystals, before the vacuum pump is turned off. If plastic slides are used, insert a thin scalpel or needle and push or twist to make an indentation. Two such marks, one at each end of the orifice, are sufficient and do not affect the sample.
5. Remove the cylinder containing the slide carriage by removing the wing nut.
6. Pull the slide carriage down so that it rests on the bottom plate, but the string to the clock mechanism remains taut.
7. Insert the slide, greased surface out and labeled end up (Fig. 8-8). The slide is inserted from above into a groove on each side of the slide carriage and slid down as far as it will go. A tension clip behind the slide holds it firmly forward.
8. Replace the cylinder with the intake orifice in front of the slide, being careful not to push the slide carriage up as the cylinder is raised into position. Replace the wing nut.
9. Mark the slide as in item No. 4. There is no provision in the machine for positioning the slide serving as the sampling surface in precisely the same vertical position on subsequent sampling periods. Therefore, it is desirable to mark the slide through the intake orifice at the beginning and end of each sampling period.
10. Loosen the wing nuts above the clock box, rotate the cover and lift it off, exposing the clock mechanism. Wind the clock. Make sure that the string wound around the cam on the front of the clock is taut as it goes down into the cylinder below where it is connected to the slide carriage. Replace the cover on the clock. The 8-day clock should be wound every 2 or 3 days to insure constant speed.
11. Remove the bolt which has been holding the vane tail stationary.
12. Turn on the vacuum pump.

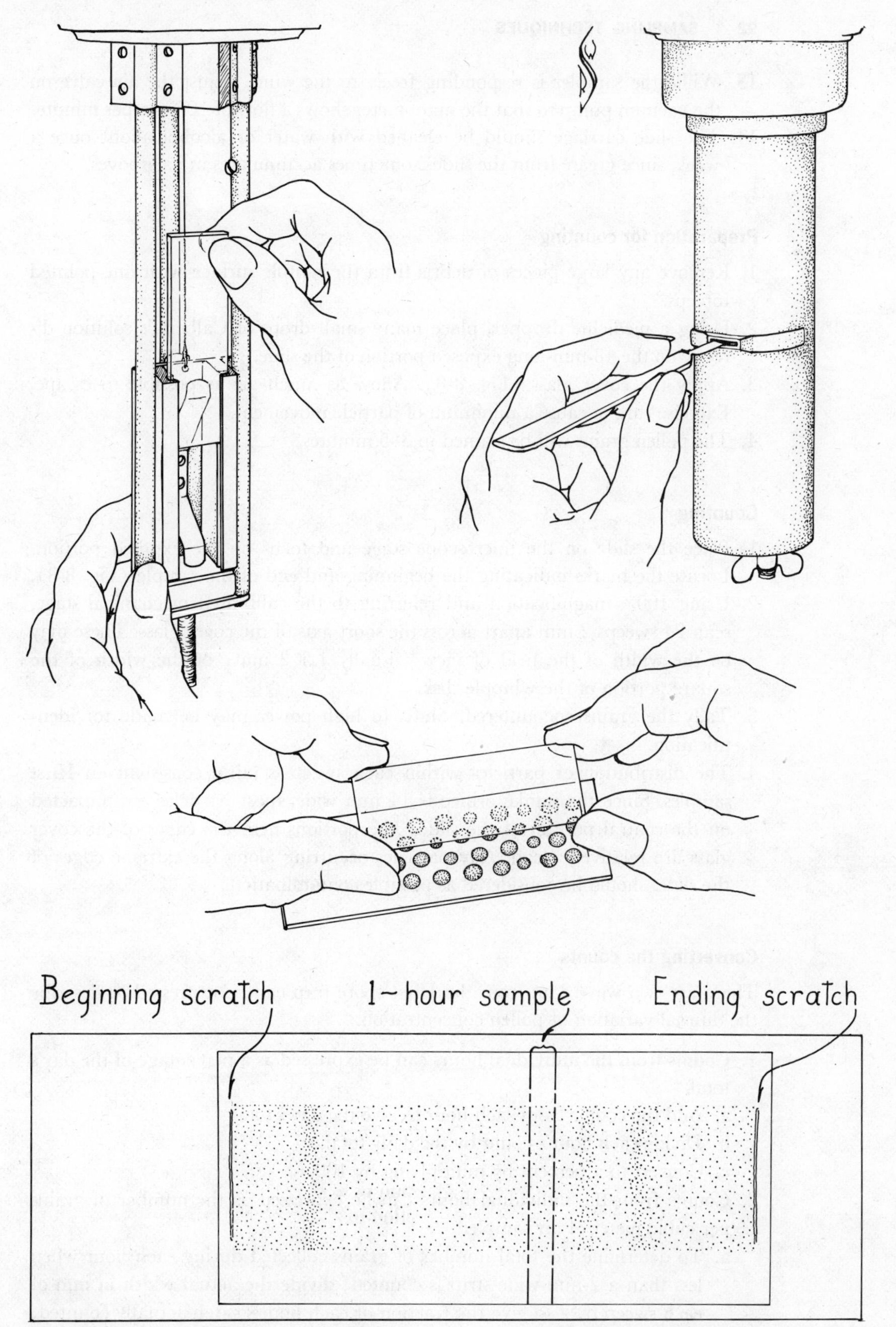

FIGURE 8-8 Loading and marking slide in Hirst spore trap and preparing the sample for analysis.

13. While the sampler is responding freely to the wind, adjust the air valve on the vacuum pump so that the manometer shows a flow of 10 liters per minute.
14. The slide carriage should be cleaned with water or alcohol about once a week, since grease from the slides sometimes accumulates in its grooves.

Preparation for counting

1. Remove any large pieces of debris from the sample surface with fine-pointed forceps.
2. Using a medicine dropper, place many small drops of Calberla's solution directly on the 48-mm-long exposed portion of the slide.
3. Apply the cover glass (Fig. 8-8). Allow as much air as possible to escape. Exercise care to cause a minimum of particle movement.
4. The pollen grains will be stained in 3–5 minutes.

Counting

1. Place the slide on the microscope stage and focus on the exposed portion. Locate the marks indicating the beginning and end of the sample (Fig. 8-8).
2. Using 100× magnification and referring to the calibrated mechanical stage, scan 24 sweeps 2 mm apart across the short axis of the cover glass. These may be the width of the field of view (usually 1.5–2 mm) or the width of the square portion of the whipple disk.
3. Tally the grains encountered. Shifts to high power may be made for identification.
4. The distribution of particles within each sweep is fairly consistent on Hirst samples. Since the intake orifice is 14 mm wide, most particles are impacted on the central portion of the slide. The portions near the edges of the cover glass are relatively clean. Any particles occurring along the extreme edges of the slides should be considered as possible contamination.

Converting the counts

There are two ways data from the Hirst spore trap can be expressed to describe the diurnal variation of pollen concentration.

1. Counts from the individual hours can be expressed as a percentage of the day's total.
   a. Add the counts from each hour to give the day's total.
   b. Divide each hour's count by the daily total.
   c. Convert to a percent by multiplying by 100.
2. Counts from the individual hours can be expressed as the number of grains per cubic meter of air sampled.
   a. To determine the total number of grains collected during each hour when less than a 2-mm-wide strip is counted; divide the actual width in mm of each sweep by 2 to give the fraction of each hour's catch actually counted; then divide the count by this fraction to give the total number of grains collected during each hour.

b. The flow rate of 10 liters/min is equal to 0.6 m^3/hour. Therefore, divide each hour's count by 0.6 to give the number of grains per cubic meter for each hour.

c. A further correction for the efficiency of the Hirst spore trap in sampling particles in the pollen size range is desirable. The efficiency will vary with particle size and wind speed as discussed in Chapters 5 and 6. Wind tunnel tests of collection efficiency for some particles have been reported by Hirst (1952). Some efficiency curves are included in the instruction booklet supplied by the manufacturer. Our wind tunnel experiments, comparing the counts with those from filters sampling isokinetically, indicate the following:

Air Speed		*Efficiency (%)*		
Meters per second	*Miles per hour*	*Timothy pollen*	*Ragweed pollen*	*Corn smut spores*
1.0	2.2	99	89	116
2.0	4.5	71	68	97
4.0	8.9	65	58	91
6.0	13.4	66	65	78
8.0	17.9	156	100	99
10.0	22.4	207	124	109

The timothy (*Phleum*) pollen in these tests was 32–35 microns in diameter, the ragweed (*Ambrosia*) pollen was 20 microns, and the corn smut (*Ustilago*) spores were 7.5 microns.

FILTER SAMPLERS

Filter samplers are not widely used for routine pollen sampling but are used for some special studies. Some samplers using filters to collect particles are discussed in Chapter 6. Filters of many sizes are made from a variety of materials and are available with a range of pore sizes. The choice of filter depends on the instrument used, the flow rate desired, and the analytical method. Pollen samples must be examined visually with a microscope. Smooth filters which retain the collected pollen on the surface and are not more than about two inches in diameter can be examined directly with a microscope. Some very large filters, or those in which the collected particles become embedded within the matrix, can be dissolved with a suitable chemical. The solution with the pollen in suspension can then be filtered through a different filter material of a size small enough to examine and which retains the pollen on its surface. With the collected pollen on a smooth filter of a size suitable to scan completely, attention must be given to the ease of identifying the pollens of interest. The filter may be opaque, translucent, or transparent and examined with reflected or transmitted light. The pollen may be examined dry or wet, uncovered or covered. It may be desirable to transparentize the filter by application of a liquid with the same refractive index as the filter material.

Identification may require that the pollen grains be stained and expanded. Information on flow rate, chemical reactions, refractive index, and physical appearance are included in catalogs supplied by the manufacturer. For pollen identification, it is desirable to choose a filter that: (1) is transparent or can be transparentized, (2) will allow one to stain and expand the pollen, and (3) permits mounting the sample in a medium having good optical properties with a high-dry objective lens. Listed below are two filter materials with a description of techniques which may be used in preparing collected pollen for microscopic examination. Discussion of these filter materials illustrates the factors which must be considered when planning a sampling program utilizing filters as the collecting surface.

Nuclepore

Nuclepore filters (General Electric Company) are transparent and therefore well-adapted to pollen sampling. Place the filter on a glass microscope slide. Apply prestained Calberla's solution or glycerin jelly and a cover glass, being careful to avoid trapping air beneath the filter. Choose a cover glass the same size or slightly larger than the filter.

Samples on large Nuclepore filters, that would require much time to scan with a microscope, may be concentrated in several ways. The Nuclepore filter may be dissolved in a solvent and drawn through a small filter of a kind not affected by the solvent.

For example, the Nuclepore filter may be dissolved in chloroform and drawn through a standard Millipore filter.

1. Working in a ventilated fume hood, dissolve the Nuclepore filter in chloroform in a filter funnel and draw the solution by vacuum through a standard Millipore filter.
2. Wash with water.
3. Remove the Millipore filter to a paper towel and allow to dry for five minutes.
4. Place four drops of ethyl cellosolve (ethylene glycol monoethyl ether) on a clean slide.
5. Place the filter on the cellosolve.
6. Allow to dry for several hours or overnight. Avoid contamination.
7. As the drying proceeds, the Millipore filter will become transparent. If any white portions remain, another drop or two of the cellosolve may be applied.
8. To stain, apply 2–3 drops of prestained melted glycerin jelly and a 1-inch round cover glass.
9. It will take several hours for the pollen to stain. Sometimes a significant portion of the grains may be poorly stained, due to coating of the pollen grains by other substances collected by the filter. Care must be taken to include these unstained grains in the count.

Molecular membrane

Molecular membrane filters are smooth, white (or black) filters which are widely used in various types of filtering operations. If simply mounted with Calberla's solution or glycerin jelly, they do not have good optical qualities at high magnification. They can be transparentized with a fluid having the same refractive index, but it may not be compatible with the staining solution. One should consult the manual supplied by the manufacturer of the filters being used.

The procedure described in the preceding paragraph can be followed for pollen collected on standard Millipore (Millipore Filter Corporation) or other similar filters starting with step #4. This should result in a mount having adequate optical properties for the use of high-dry magnification and having stained and expanded pollen grains.

References

Brown, C.A. 1960. Palynological techniques. Published by the author, Baton Rouge, La. 188 p.

Conn, H.J. 1961. Biological stains. 7th ed. Williams & Wilkins Co., Baltimore, Md. 355 p.

Dingle, A.N. 1957. Hay fever pollen counts and some weather effects. Bull. Amer. Meteorol. Soc. 38:465–469.

Durham, O.C. 1946. The volumetric incidence of airborne allergens. IV. A proposed standard method of gravity sampling, counting and volumetric interpolation of results. J. Allergy 17:79–86.

Gray, P. 1964. Handbook of basic microtechnique. 3rd ed. McGraw-Hill Book Co., New York. 302 p.

Gurr, E. 1962. Staining, practical and theoretical. Williams & Wilkins Co., Baltimore, Md. 631 p.

Harrington, J.B., G.C. Gill, and B.R. Warr. 1959. High efficiency pollen samplers for use in clinical allergy. J. Allergy 30:357–375.

Hartley, W.G. 1964. How to use a microscope. Natural History Press, Doubleday & Co., New York. 255 p.

Hayes, J.V. 1969. Comparison of the rotoslide and Durham samplers in a survey of airborne pollen. Ann. Allergy 27:575–584.

Hirst, J.M. 1952. An automatic volumetric spore trap. Ann. Appl. Biol. 39:257–265.

Möllring, F.K. [1968]. Microscopy from the very beginning. Carl Zeiss, New York. 63 p.

Ogden, E.C., and G.S. Raynor. 1960. Field evaluation of ragweed pollen samplers. J. Allergy 31:307–316.

Ogden, E.C., and G.S. Raynor. 1967. A new sampler for airborne pollen: the rotoslide. J. Allergy 40:1–11.

Ogden, E.C., J.V. Hayes, and G.S. Raynor. 1969. Diurnal patterns of pollen emission in *Ambrosia, Phleum, Zea,* and *Ricinus*. Amer. J. Bot. 56:16–21.

Raynor, G.S., and E.C. Ogden. 1970. The swing-shield: an improved shielding device for the intermittent rotoslide sampler. J. Allergy 45:329–332.

Sass, J.E. 1958. Botanical microtechnique. 3rd ed. Iowa State College Press, Ames, Iowa. 228 p.

Solomon, W.R., A.W. Stohrer, and J.A. Gilliam. 1968. The "fly-shield" rotobar: a simplified impaction sampler with motion-regulated shielding. J. Allergy 41:290–296.

Venning, F.D. 1954. Manual of advanced plant microtechnique. W.C. Brown Co., Dubuque, Iowa. 96 p.

Wodehouse, R.P. 1935. Pollen grains. McGraw-Hill Book Co., New York. 574 p. (Reprinted 1959 by Hafner Publ. Co., New York).

9
IDENTIFICATION OF POLLEN

Obviously, one of the most important activities in sampling airborne pollen is identification of the particles of interest that are found in the samples. The most careful attention to proper techniques for obtaining the samples and preparing them for analysis is of little avail if determinations are faulty. Even if only ragweed pollen is to be recorded, care must be taken to distinguish these grains from others of similar size and having similar markings. Most pollen grains that become airborne in sufficient quantity to be of importance may be identified with confidence after a period of practice with reference slides of known pollens and with available publications.

REFERENCE MATERIAL

The intended scope of a particular study will determine the selection of reference materials. In some studies, the number of and distribution of fairly distinctive pollens is the only information desired and precise identifications are not required. The materials discussed here assume the necessity for the distinction of several types and this will require recognition of pollen features as usually revealed at a magnification of 400 diameters.

It is wise to gather and become familiar with basic materials well in advance of the field sampling season. These materials will include: (1) publications dealing with pollen morphology and identification and also those with information on distribution, abundance, and flowering times of plants that produce airborne pollen, (2) reference microscope slides for comparison with field-collected pollen (see Chapter 4), (3) photomicrographs and drawings of pollen, and (4) laboratory equipment and supplies (see Chapter 8).

Of the many publications that help in the identification of pollen, a few are singled out here as being especially useful. These and others are referenced at the end of the Chapter.

Pollen grains (Wodehouse 1935). This pioneer study of airborne pollen is one of the most useful for the beginner. The discussions of evolutionary aspects of pollen form and the effects of their developmental environment on their mature appearance help in understanding the morphological characters. Keys and pollen descriptions of genera and species are accompanied by excellent drawings. The drawings show overall form as interpreted from several levels of microscope focus and should be studied along with reference slides and photographs. Some of the species described are not commonly airborne but indicate similarity or diversity of pollen form in closely related plants.

Atmospheric pollen (Wodehouse 1942). The same type of drawings are used as in the previous reference, but pollens illustrated are limited to "all species . . . known to the author to have been caught in the atmosphere and others which might reasonably be expected to be caught. . . ." This includes a few spores of ferns and fern allies and some pollens which are generally insect-pollinated. While the total is less, several pollens not illustrated in the previous work are shown. These drawings are accompanied by brief descriptions. A useful feature is a listing of "similar grains" following many of these descriptions.

Hayfever plants (Wodehouse 1945, revised 1971). This contains the same sort of drawings shown in the previous publications and mostly of the same species. References for the identification of plants and tables showing times of various pollen occurrences are given for ten regions of the conterminous United States. The revised edition has added new material.

Pollens and the plants that produce them (Solomon et al. 1967). Photographs and drawings of selected pollens and photographs of some of these plants in flower are shown. Important hayfever plants and their distribution in the United States are briefly mentioned. Fewer pollens are illustrated than in the Wodehouse texts, but the photographs with marked points of reference may be more directly comparable to grains as seen through the microscope. Terminology is simple and generally nontechnical. Comparisons are often made with easily associated objects such as "two fried eggs in a pan," "flying saucers," and "golf ball-like."

An atlas of airborne pollen grains (Hyde and Adams 1958). This deals with airborne pollen grains in Great Britain. There are photographs of ninety-two species, with each usually at two levels of microscope focus and often oriented in different ways. Uniform, technical descriptions employ a minimum of palynological terminology, all of which is adequately defined in the glossary. Many of the European pollens illustrated here have been introduced into the United States as ornamental or weedy plants and others have pollens similar to closely related native species. The photographs appear much the same as fresh pollen viewed under the microscope in routine examinations. The atlas can be used for direct comparison of field-collected unknown pollens, but the worker should have knowledge of the same or similar plants growing in the area and should have an adequate collection of reference microscope slides of these for comparison.

Other texts designed primarily for students of fossil pollen are also useful. Fossil pollens are devoid of cell contents and the innermost wall. Classification is generally more sophisticated than in texts dealing exclusively with airborne modern pollens and subtleties of sculpture and wall structure are stressed. The worker should keep in mind that sizes and shapes may not be the same for pollen of the same species in fresh and fossil condition. Following is a brief account of some of these references.

An introduction to pollen analysis (Erdtman 1943). This contains drawings and descriptions of pollen grains and fern spores of wide distribution, but with emphasis on those found in northwestern Europe and the United States.

Pollen morphology and plant taxonomy: angiosperms (Erdtman 1952). This is an extensive treatment of pollens of worldwide distribution with carefully executed idealized drawings of over 600 species and descriptions of many more.

Families having similar pollen grains are listed. This is not a general reference for the beginning worker in airborne pollen studies but can be useful in many ways. Later volumes in this series (1957, 1965) cover the mosses, ferns, and conifers.

Textbook of pollen analysis (Faegri and Iversen 1964). Basic pollen terms are defined. There is a key to types of pollen in northwestern Europe. There are special keys for sedges (Cyperaceae), grasses (Gramineae), and plantains (Plantaginaceae). The main key can be used, with reservations, for pollen types found in the United States.

How to know pollen and spores (Kapp 1969). Many kinds of pollen found in the United States (including some not illustrated in other texts) are shown in drawings. General distribution ranges are given. Other "palynomorphs": algae and spores of fungi, mosses, ferns, and fern allies which might be found in air samples are also described.

Morphologic encyclopedia of palynology (Kremp 1965). This volume helps to equate the many synonomous terms encountered in palynology.

Grana (and its predecessor, *Grana Palynologica*), *Pollen et Spores,* and *Review of Palaeobotany and Palynology* are three periodicals which publish wide-ranging articles in all fields of palynology. *Pollen et Spores* also publishes a supplemental bibliography of considerable value in keeping workers informed of current articles in other journals.

Some publications indirectly aid in the identification of pollen by showing, through maps or tables, the time and abundance for many pollens in various regions of the United States. These include: the *Statistical report of the Pollen and Mold Committee of the American Academy of Allergy* published for the Committee by Ross Laboratories, *The what, when, where of hay fever* published by Abbott Laboratories, and *Regional pollen guide* published by Hollister-Stier Laboratories.

In addition to these general references, several publications include photographs of pollen which may prove useful. Scanning electron micrographs may not have the appearance of pollen grains as seen in routine examinations, but they do show surface details that can only be inferred from observations made with the ordinary light microscope. Comparisons of these with reference slides aid in the interpretation of pollen characters. A personal collection of photographs is helpful in building an organized file of pollen information (Samuelsson 1965). A punched card system may be adapted for easy retrieval of pollen data (Faegri and Iversen 1964: 199). A camera attachment, especially one for making Polaroid prints is valuable. A single pollen grain may be photographed at several levels of focus, and differences due to these various views may be more apparent than by direct microscopic observations. The photographs may be suitably backed and used as "flash cards" in pollen recognition exercises.

Drawings of pollen, either freehand or with the aid of some mechanical device, help to focus the worker's attention on the outstanding features of pollen form.

POLLEN MORPHOLOGY

With the aid of electron microscopy, the study of the developmental morphology of pollen has become an end in itself. Interesting summaries of these studies are found in Echlin (1968), Dahl (1969), Gullvåg (1966), and Heslop-Harrison

(1968). However, only a cursory knowledge of morphology is required for basic pollen identifications.

Certain tissues give rise to pollen mother cells. These are surrounded by the tapetum, a layer of nonsporogenous tissue commonly one cell thick. All of these tissues are surrounded by the anther wall. During the developmental period, some of the events influencing mature pollen form are: (1) insulation of the pollen mother cell from others in the pollen chamber by a callose sheath and the reduction division of the pollen mother cell with the eventual formation of four pollen grains (tetrad), (2) development of the rudiments of exine (outer wall) and apertures (wall thinnings), dissolution of the callose wall and release of the individual grains into the pollen chamber, and (3) elaboration of exine (perhaps by tapetal nourishment) and formation of the intine (innermost wall). The apertures, which may be rounded pores or elongated furrows or a combination of these, are generally marked by a thinning or absence of the outer wall layer. The intine is continuous beneath these structures and is frequently considerably thickened. This developmental sequence is the subject of continuing intensive studies and may be modified as a result (Christensen et al. 1972).

The identification of fresh pollen is concerned mostly with germinal apertures, sculpture, size, shape, color, and cell contents. This order of listing is approximately their order of importance.

Germinal apertures

In most, if not all, comprehensive studies where keys for the identification of pollen are employed, the structure of the keys is based on the form and number of germinal apertures (Wodehouse 1935, Faegri and Iversen 1964, Hyde and Adams 1958, Kapp 1969). However, to the beginning worker, it is likely that the characters of size, staining intensity, and gross sculpturing will be more impressive at the low magnifications usually employed.

The position of the developing pollen grains in the tetrad and, in the broadest sense, the ancestry of the plants which produce them are often predictors of the numbers of apertures found in the mature grains. In monocots, the common configuration of the four grains is a square; in dicots, the tetrahedral tetrad is most common. Rhomboidal, linear, and other arrangements are also possible (Fig. 9-1). The individual pollen grains are generally spheroidal and designation of specific areas follows such global terminology as polar, meridional, and equatorial. Lines from the center of square or tetrahedral tetrads, for instance, can be drawn to the center of the outer face of the individual grains. These lines can be considered the polar axes of the grains. The center of the tetrad is the proximal portion, and the proximal pole of the individual pollen grain is located at this intersection. The point on the opposite side of the grain is the distal pole (Fig. 9-1A).

Germinal apertures of some sort are formed in most pollen grains, although inaperturate grains occur in the gymnosperms, monocots, and dicots. These are usually thin-walled pollens. In a common monocot form as shown in the stylized square tetrad in figure 9-1A, an aperture is formed at the distal pole. The form and size of the aperture varies with the plant's ancestry, although it is usually consistent within families and often over broader relationships. The pollen wall

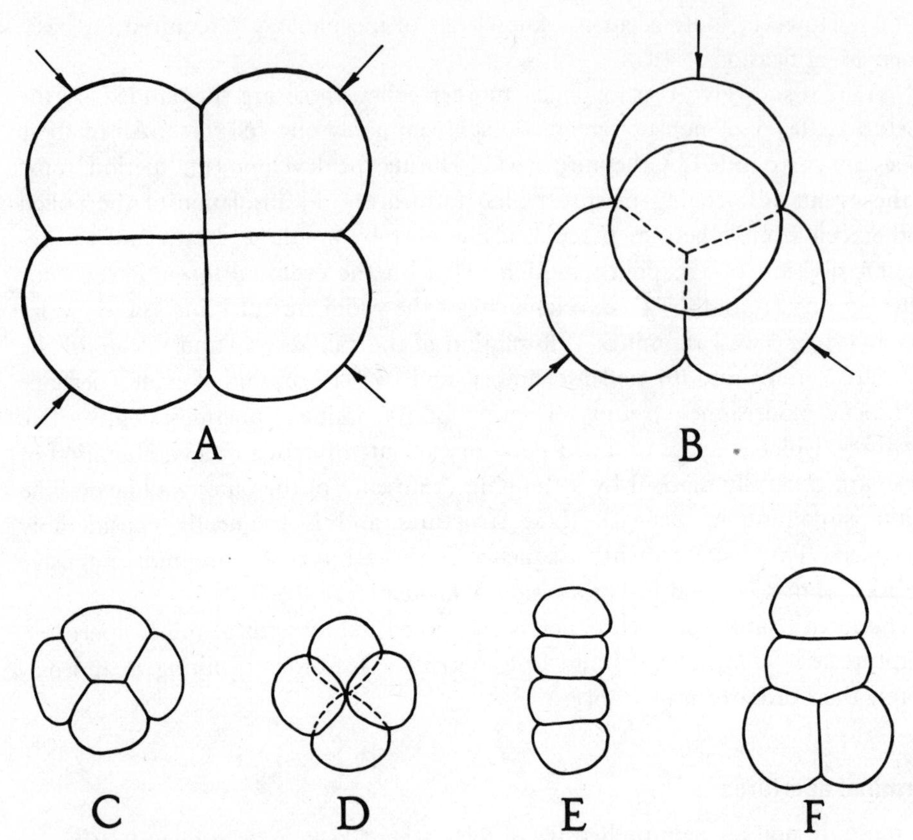

FIGURE 9-1 Pollen tetrads. A, square tetrad. Arrows indicate positions of the distal poles of individual grains. The proximal poles are at the common center. B, tetrahedral tetrad. The central grain is closest to the viewer. Dashed lines indicate the proximal portion where the grains meet at a lower level of focus. The polar axis of the central grain is approximately along the line of sight from its center to the center of the tetrad. The approximate positions of the distal poles of the other three grains are indicated by arrows. C, rhomboidal tetrad. D, decussate or cross tetrad. E, linear tetrad. F, T-shaped tetrad. All of these tetrads are from mature pollen of Typha latifolia.

at the germinal aperture is structurally modified, allows an exit for the pollen tube, and permits expansion of moistened grains.

In the tetrahedral tetrads commonly found in dicotyledonous plants, each grain has an equal contact with the other three grains (Fig. 9-1B). The position of germinal apertures is probably related to these contacts. Apertures are not formed at the distal pole as with the monocots.

Apertures may be furrows, pores, or a combination of these. The distinction between furrows and pores is arbitrary. Pores are nearly circular. Apertures are called furrows if their long axes are more than twice their width. A plane midway between the poles and perpendicular to the polar axis intersects the outer surface of the pollen grain at the equator, usually the place of greatest diameter along the polar axis. With pollens in polar view as determined by the convergence of their apertures toward a point in the center of the grain (pole), the equator is

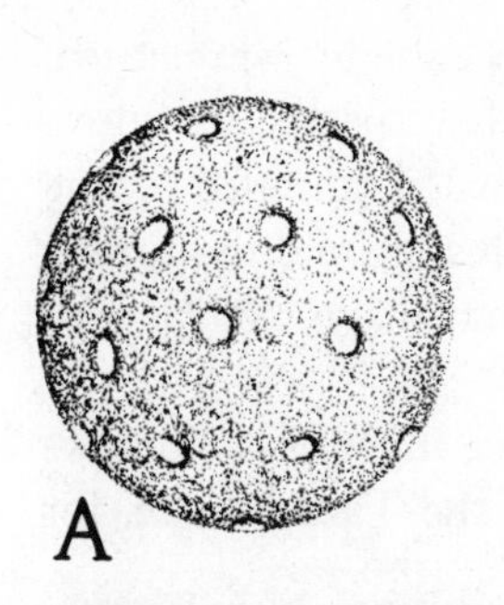
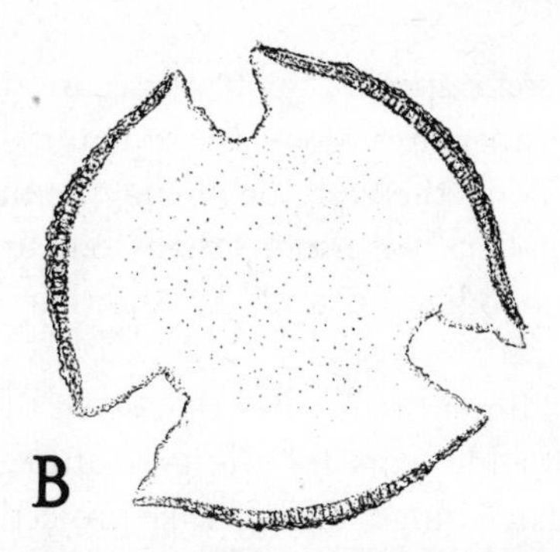
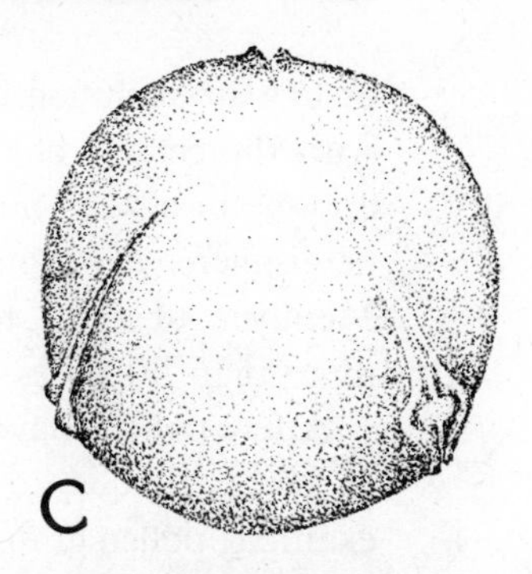

FIGURE 9-2 Kinds of germinal apertures. A, porate pollen of Amaranthus. B, furrowed pollen of Quercus in polar view (optical cross section). C, furrows and associated pores of Nyssa in oblique polar view.

essentially the outline of the grain. For practical purposes, the outline of circular grains in which polar areas cannot be determined is also the equator even though this may not be true in the developmental sense. Furrows in the pollen of dicotyledonous plants are generally three, equally spaced around the grain with their long axes directed toward the poles and are accordingly termed meridional furrows. If grains have three free pores or three pores enclosed by furrows, these are usually at the plane of the equator and are equally spaced (Fig. 9-2).

These comments concerning common aperture configuration in the typical gymnosperm-monocot form and the dicot form are generalizations and numbers and positions of apertures vary considerably in each. As a practical consideration in studies of airborne pollen, however, all pollens with single apertures are from gymnosperms or monocots; all pollens with three apertures are from dicots. Of the genera listed in Chapter 3, only one or possibly two might seem to be exceptions.

Furrows or pores may be of definite outline or may be ragged in appearance. They may also have distinct margins, often the result of wall thickening in a narrow zone. For example, the birch family and members of other closely related families typically have "shield-shaped" thickenings, named aspides by Wodehouse (1935), surrounding their pores (Fig. 9-3A).

Apertures also may be compound structures with furrows enclosing pores or overlying shorter furrows. The inner furrow may be transverse (Fig. 9-3B), having its long axis oriented at right angles to the long axis of the meridional furrows. Some apertural arrangements become complex and considerable experience is necessary to interpret their structure correctly.

In dicotyledonous plants, the usual number of apertures is three, although four

FIGURE 9-3 A, Ostrya pollen showing the three pores with aspides and onci in polar view (optical cross section). B, pollen of Angelica in the Umbelliferae showing the elongate meridional furrow and the smaller, transverse aperture in equatorial view.

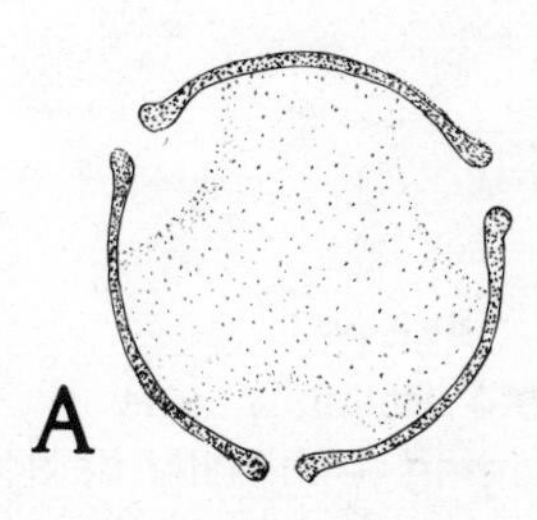
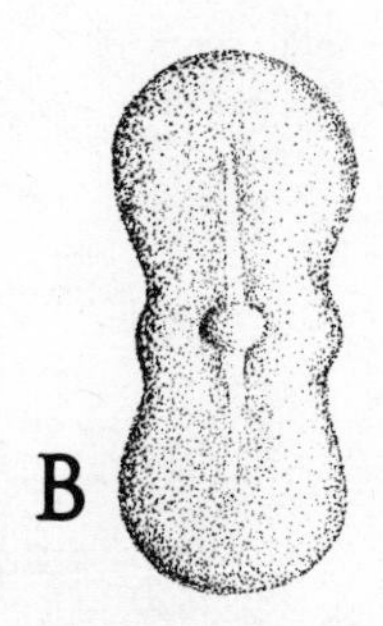

or five are common in some species. Such grains are usually radially symmetrical with the centers of the apertures along the equator. In a few species, apertures may be in a zone not along the equator or may be more-or-less confined to one hemisphere. Airborne pollens with more than four apertures are mostly porate. Members of the Chenopodiaceae and Amaranthaceae may have pollen with more than 50 pores.

A few pollens have different structures that serve the same function as furrows or pores; that is, to provide exits for the germ tubes. In the Taxodiaceae, for example, pollen grains have bulges or fingerlike projections.

A covering of the apertural membrane called the operculum is sometimes found. Samples of airborne pollen are usually stained in basic fuchsin and mounted in an aqueous medium. In such samples, an operculum is distinct. It stains and contrasts with the inner wall layers on which it rests. The operculum of the grasses in these preparations is often ephemeral. It may appear as a half-raised lid or be lost altogether. The operculum of *Plantago lanceolata* seems surrounded by a moat (the nonstaining inner wall) evoking the analogy to its "doughnut" appearance (Fig. 9-4). Scattered processes and remnants of the outer wall frequently seen on aperture membranes are probably not opercula in the sense of the definition of Wodehouse (1935), but are important aids in pollen identification (e.g., *Platanus*).

Sculpture

Many pollen grains are strikingly ornate with various processes and shapes which are enhanced by proper staining. The more spectacular pollens are from plants which are normally insect-pollinated, but common airborne grains also show diverse patterning.

The outer wall (exine) may be distinctly ornamented with striations, netlike patterns, spines, or other processes on its surface. It may also show differences in internal construction. The surface features are usually referred to as sculpture, the internal features as structure, but the technical differences are more complex. The cell contents of fresh pollen often obscure subtle structural details, therefore

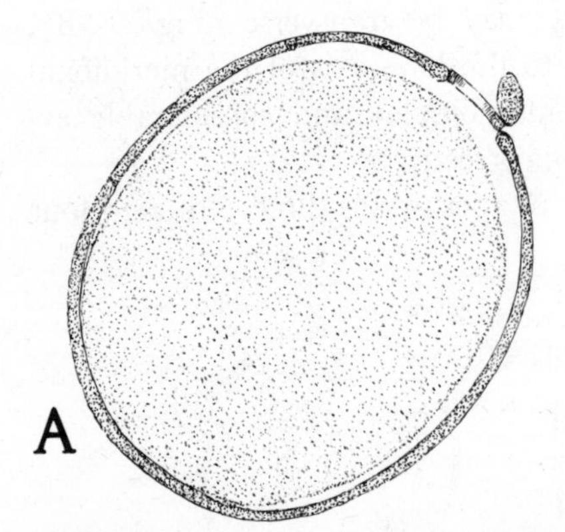

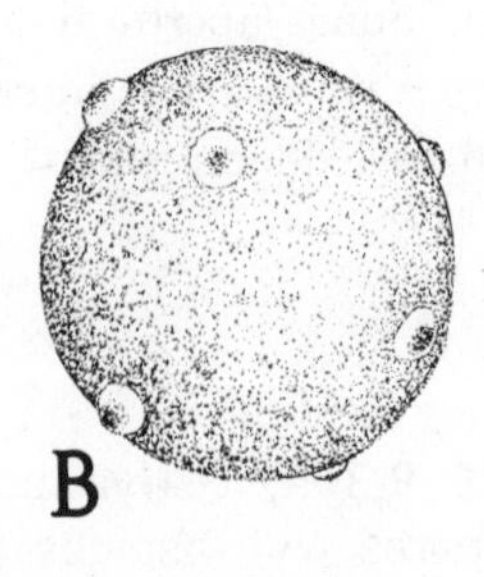

FIGURE 9-4 Pollen grains showing opercula. A, grass (Gramineae) pollen showing a partially detached operculum (optical cross section). B, Plantago lanceolata showing opercula in surface view.

structure is relatively subordinate in the study of these pollens. Structural characters which can be discerned will be considered with sculpture here.

Electron microscope studies have greatly increased our knowledge of pollen development, but have also promoted controversy as to the sequence of developmental events, the function of the various involved tissues, and the systematic value of observed differences in fine details. It is generally agreed that the first formed portion of the wall which is recognizable in mature pollen is a layer of rodlike elements perpendicular to the surface of the cell. The outer portions of these may spread and coalesce in a roof, varying in completeness in different pollen types. The bases of these rods may similarly merge to form a foot layer. Beneath the foot layer, an inner layer of different developmental organization and staining reaction is formed. This inner layer appears homogeneous by light microscopy and is made up of the same material as the outer layers. This material is sporopollenin which is very resistant to chemical attack and is responsible for pollen wall preservation. In addition to roof, rod, and foot layers, there may also be processes on the roof layer which form the sculpture elements which are of great importance in distinguishing the different pollen types (Fig. 9-5A). Some of these are described as spines, warts, pits, and many other terms whose technical characteristics are described in textbooks of pollen morphology (Fig. 9-5). Any of these outer layers may be modified. The roof may be missing, for example, resulting in mature pollen with sculpture consisting of the rod layer with their ends nearly free. This will appear similar to spinules imposed on the roof of another type of pollen. Such differences in sculpture are important in pollen identification.

In specimens stained with basic fuchsin, the outer layers of the exine will stain deep red; that beneath the foot layer will stain pink. The internal layer, the intine, does not stain with basic fuchsin. It is cellulosic and does not contain sporopollenin.

So, the wall of pollen has many characters useful in pollen identification. With airborne pollen, the tendency of cell contents to obscure internal details of the wall makes sculpture elements (spines, etc.) relatively more important. But

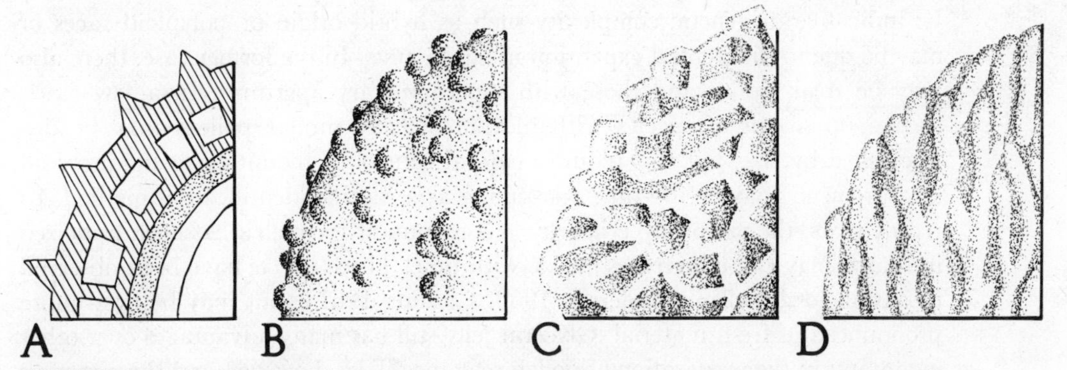

FIGURE 9-5 Sketch to show exine cross section and surface markings. A, stylized cross section of a pollen wall showing the dark-staining outer wall including spines, roof layer, rod layer, and foot layer; the lighter staining layer beneath this (indicated by stippling); and the nonstaining intine. B, granular or warted pollen surface. C, netted pollen surface. D, striate pollen surface.

staining reaction, differences in thickness of the various layers, and their complexity (particularly the rod layer) are also important. Characters of the intine are of importance in only a few airborne pollens. Those with thin outer walls commonly have thick intines. Some pollens have much thickened intines associated with the apertures.

Textbook descriptions of the structure and sculpture of pollen walls are precise, with each wall layer and sculpturing process being strictly defined and named. It is necessary to study the standard works (Erdtman 1943 and 1952, Faegri and Iversen 1964, Kremp 1965) to properly classify these features. In the following remarks, pollens are considered simply (if not precisely) to be smooth, granular, striate, netted, or spiny.

1. Smooth grains are those with sculpture elements generally less than one micron in their greatest dimension and not organized in any distinct pattern.
2. Granular grains have slight excrescences not much greater than one micron. These may be described as knoblike, wartlike, etc., but they are generally not grouped in an organized manner.
3. Striate grains are distinctly streaked by ridges, grooves, or rows of granules. The striae are generally parallel and may be straight or whorled.
4. Netted grains have elements in rows which interconnect, forming netlike meshes. These may be ridges, grooves, or processes; or they may be due to subsurface features.
5. Spiny grains have generally pointed processes distinctly larger than granules but, in our broad definition, they include grains with reduced spines but of pollen morphology similar to closely related plants.

Size

Among the smallest pollen grains that are commonly airborne is *Urtica dioica,* about 10 to 12 microns, while some species of *Abies* may be as large as 125 microns. The size of the pollen of a particular species usually falls within a predictable range of a few microns. With most plants, much divergence from normal size might lead to suspicion of the mounting techniques employed or to the accuracy of labeling. A considerable range in pollen size within a mount may be indicative of genetic complexity such as hybrid origin or polyploid races or may be due to differential expansion in old mounts. In the former case, there also may be freak grains and those with supernumerary apertures. In a few cases, species or groups of species with morphologically similar pollens may be distinguished by size, but this requires careful control of mounting techniques and measurements may be too time consuming as a routine identification method. In recent years, the use of glycerin jelly as a mountant for pollen has been criticized in that it may cause increases in size with time. These effects have been observed mostly in fossil pollen (Andersen 1960, Cushing 1961), but may be even more pronounced in fresh material. Glycerin jelly still has many advantages over other mountants in the preparation of pollen reference slides, however, and the materials recommended as substitute mounting media are, for various reasons, unsuitable for fresh material. To help prevent the increase in size, the glycerin jelly or slide should not be too hot (Ting 1966); the coverglass should not be pressed down to provide an even mount; and the coverglass should be ringed with a material such as paraffin which will support it and prevent drying of the glycerin jelly.

See Traverse (1965) for detailed techniques. It is necessary to compromise on thickness of mounts: thick enough so grains are not distorted and thin enough so oil immersion objectives can be used. Measurements should be made and recorded on the label as soon after mounting as possible. Excessive expansion of pollen usually requires at least a few days in unringed mounts, but it is sometimes almost immediate. With experience, it is usually possible to determine if grains with which the worker is familiar have expanded with time. Reserve pollen, dried and stored in packets (see Chapter 4), can be used to make replacement reference slides. With other slides of doubtful quality (especially those which have been stored for some time), size checks may be made by comparing them with temporary mounts in aqueous media.

In spite of drawbacks, size is used considerably in routine pollen examinations. It may be the single most important characteristic for the rapid identification of very small or very large pollens, but it is less helpful in determining the large number of pollens that fall between 20 and 40 microns.

In radially symmetrical spheroidal grains in polar view, there is often only one meaningful measurement to be made (e.g., the equatorial diameter of members of the Betulaceae). On grains of unusual shape, several measurements may be taken and ratios of dimensions calculated (e.g., the length, breadth, and depth of winged conifer pollen compared with the size of the wings). The measurements should be clearly described. For example, spiny grains are usually measured to the bases rather than the tips of the spines. Ambiguous descriptions of these dimensions result in worthless data. The dimensions measured for some pollen types have become more-or-less standard, and it is helpful for comparative purposes to follow these conventions.

The relationships between various measurements may have diagnostic value. Shape is described as the ratio between the length of the polar axis and the equatorial diameter for grains appropriately oriented. Another useful measurement which serves to separate similar grains is the polar area index. This has been variously described, but is essentially the distance between the apices of germinal apertures (or their margins) relative to the equatorial diameter of the grain.

Shape

The shapes of airborne pollens have often been described as mostly spheroidal. This will vary with the type of pollen and the adhesive and mountant employed. Pollens with long furrows often increase greatly in volume when wetted, with the elasticity of the furrow membrane making such expansion possible. Grains with three such furrows are often oriented in equatorial view. The outline of these unexpanded grains is ellipsoidal or "boat-shaped." When the grain is wetted, it expands almost instantly and the sudden increase in volume causes it to tear loose from the adhesive surface, spin rapidly, and come to rest in polar position. In this view, the grain appears spheroidal or, if the three furrow membranes bulge greatly, subtriangular. If the pollen grain remains in equatorial orientation, it will increase in equatorial diameter while the polar axis is shortened. The shape of the grain may, therefore, change from ellipsoidal to spheroidal or even be flattened at the poles (Fig. 9-6). The normal expansion and orientation of pollen grains may be affected by the mounting medium and the adhesive used to trap the air-spora. For example, pollens with long, functional furrows buried

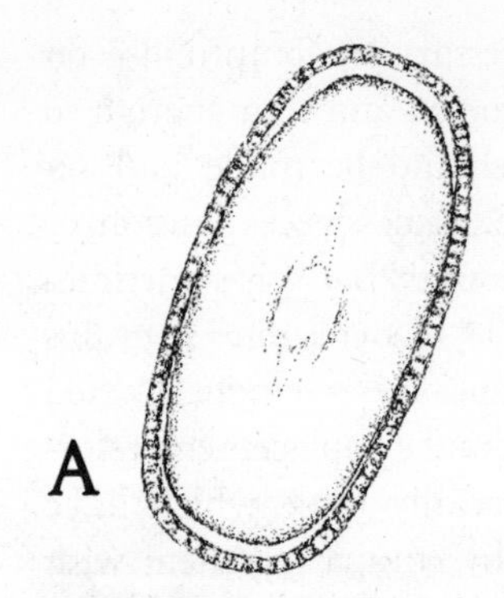

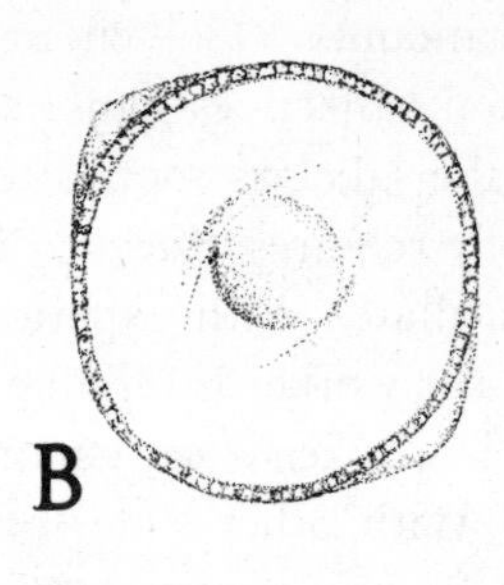

FIGURE 9-6 Change in shape of Ailanthus pollen in different mounting media. A, unexpanded in silicone oil. B, expanded in glycerin jelly. Both grains are in equatorial view.

in a too thick layer of grease are seen as ellipsoidal forms in equatorial or oblique equatorial views. These are not wetted by the mounting medium and, consequently, their appearance does not change. Other adhesives may allow contact with the mounting medium, but hold the grain so firmly that the usual change in orientation (from equatorial to polar view) does not take place. The age and condition of pollen may also affect expansion. Some pollens may have thick walls and relatively nonfunctional apertures which do not accommodate changes in volume, while others have internal structures which allow swelling of cell contents with little outward change in the appearance of the pollen grains (Skvarla and Larson 1965). Thin-walled grains without such modifications or germinal apertures may rupture on wetting.

Shape has been exploited in pollen identification, especially by those working with fossil pollen. This character has been less employed by those working with contemporary pollen, probably because of the tendency of many such pollens to be oriented in polar view upon mounting in aqueous media. Shapes of pollen in equatorial and polar view are described by Erdtman (1952). For grains in equatorial view, nine shape classes are listed from those in which grains are much flattened at the poles to those in which grains are elongate with the polar axis more than twice the equatorial diameter. For grains with three apertures and oriented in polar position, six general outlines are shown. These shape characterizations concern grains without cell contents. The shapes of the same types of grains in fresh condition in aqueous mounting media will likely be modified.

There may be other variations in shape caused by appendages such as the bladdery "wings" of the pines and their relatives or the germinal papilla of *Sequoia*. In some plants, the individual microspores may not separate and the mature pollen may be released from the anther as tetrads or as compound grains in multiples of four cells. Most of these types are insect-pollinated, but tetrads of the common cattail (*Typha latifolia*) and compound grains of *Acacia* and closely related plants may be encountered locally.

Color

Natural color of fresh, airborne pollen is generally some shade of yellow to nearly colorless. It is not normally distinctive and is consequently of little use in pollen identification. Partly for this reason, field-collected pollen is usually stained and mounted in an aqueous medium for optimum contrast in the study of other pollen characteristics. Some insect-pollinated plants have more colorful pollen with reports of blue, green, red, and black (Cranwell 1953, Hodges 1952, Reiter 1947). Individual pollen grains under the microscope may appear to be a different color than the same pollen in mass, such as that packed by bees. Some of the unusual colors of entomophilous pollen are due to oily droplets on their surface. It has been suggested that these substances may serve as insect attractants or to protect the pollen from ultaviolet radiation (Heslop-Harrison 1968). The pollen of some species, not normally airborne, is bound together by this material, dispersed in clumps, and occasionally trapped in significant amounts on air samples (e.g., *Taraxacum*).

Airborne pollen is usually stained with basic fuchsin. Those with exines of average thickness stain magenta; those with thinner exines are pink or lavender. These differences in staining intensity may afford a crude separation of disparate pollen types. Lightly stained grains of the thin-walled *Morus* or *Urtica,* for example, contrast strongly with more deeply stained pollen collected at the same time. Reference slides of pollen stained with basic fuchsin and mounted in glycerin jelly have a tendency to change color over the years with blues eventually predominating.

Pollen is sometimes dyed by water soluble pigments contained in other substances trapped along with the grains. These may impart an orange-red stain.

Old, presumably refloated, pollen may exhibit an olive to brown pigmentation of cell contents. In some cases, old pollen may show negative staining with cell contents accepting the basic fuchsin dye while the exine remains unstained. Fluorescence of pollen is common and grains have been stained with fluorescing dyes (Shellhorn et al. 1964). As yet, these methods have not aided in pollen determinations.

Cell contents

The appearance of the cell contents and the inner wall (intine) may occasionally aid in pollen identification. For example, the characteristic appearance of *Rumex* pollen has been attributed to the inclusion of spherical starch bodies. The outer wall of *Rumex* bears features which would make identification quite simple, but these are not easily seen without optimum staining, and the appearance of the cell contents becomes the most outstanding feature. Other cell inclusions in pollen (e.g., *Larix, Chamaecyparis,* Juncaceae, *Quercus,* and others) have been described by a number of authors as "rounded or elongated granules," "hyaline plugs," etc.

The average thickness of the intine of most medium-sized pollen grains is probably about one micron. Unusually thick intines are found in a number of airborne pollens and may be diagnostically important. These grains frequently have thin, relatively featureless exines, and the intine is often irregularly thickened. This condition is striking in the Cupressaceae, for instance, where the interface of the intine and cell contents is described as star-shaped in optical cross section. The major portion of these grains is composed of intine. Other pollens with distinctive intines are in the Cyperaceae where the intine is frequently much thickened at the apex of the pear-shaped grains and in *Populus* where the fairly thick intine may separate from the delicate exine. Cyperaceous grains which tend to be more spherical and whose exines are crumpled may occasionally mimic *Populus* pollen. The intine beneath the pores of aspidate grains (e.g., Betulaceae) is thickened. This feature has been termed the oncus by Hyde (1954). See Fig. 9-3A.

Methyl green and eosin may be employed for contrast staining of exine and cell contents (Wodehouse 1935). Other dyes may be used for the same effect and are helpful in studying characters of intine and cell contents, but basic fuchsin probably is still preferable for routine examination of field-collected samples.

CHARACTERS OF COMMON AIRBORNE POLLENS

The outstanding characters of most of the pollens of the genera listed in Chapter 3 are charted on the following pages. The characters are of a general nature with inaperturate types designated as aperture number 0. Aperture types include porate (with pores), colpate (with furrows), and colporate (with furrows and secondary pores or furrows). Pollens with unusual apertures are not categorized, but are explained under the discussion of the group. Size refers to the typical range in size for the genus as represented in the conterminous United States. Line drawings give an impression of appearance, including size, of the various pollen types. These are not accurate representations, but emphasize one or two outstanding characters for each pollen shown. Accordingly, some are depicted in surface view and others in cross section; some are shown in polar and some in equatorial view. The views selected are supposedly those most commonly seen for field-collected, airborne pollen.

The charts show eight "pollen groups." These are broad, very loosely defined categories meant to provide a basis for discussion of features useful in separating similar grains. A single pollen type might logically be included in two or three of these groups. The following discussions are designed for the inexperienced pollen worker to indicate diversity of pollen form without unduly complicating the subject. Consequently, some grains may be included in categories where they do not strictly belong.

The order of discussion is one of convenience, but is based to some extent on the expected abundance of pollens on field-collected samples:

1. Spiny grains
2. Winged grains
3. *Juniperus*-type grains
4. Other spheroidal, inaperturate grains
5. Porate grains
6. Netted grains
7. Other triaperturate grains
8. Irregularly shaped grains

1. Spiny grains

Airborne pollens include only a few genera that have evident spines. The most common spiny types encountered are of the family Compositae. Spiny grains may be roughly grouped into four categories: (1) short, conical spines and short furrows, (2) short spines and long furrows, (3) longer, sharp-pointed spines and long furrows, and (4) those with spines borne on ridges.

Short, conical spines and short furrows

This would include *Ambrosia* (Figs. 9-7 and 9-8), *Iva* (except *I. xanthifolia*),

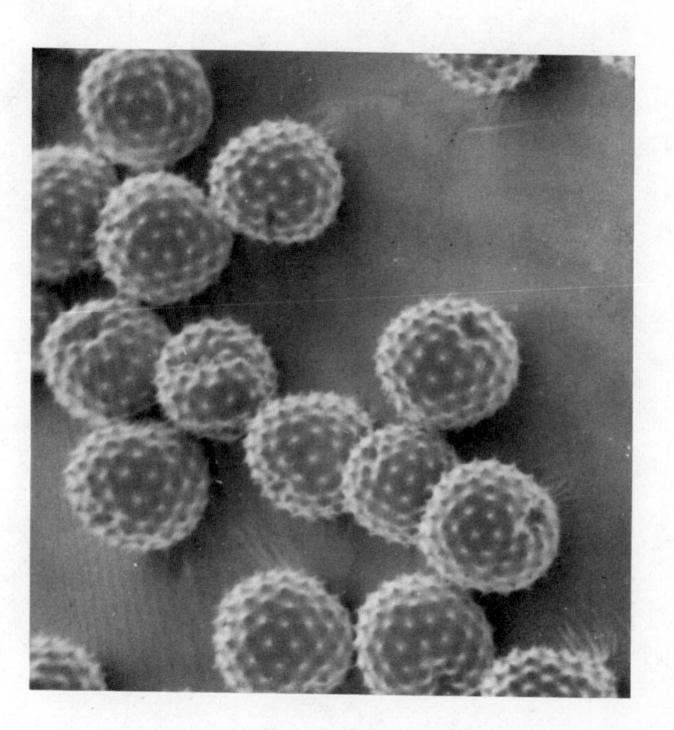

FIGURE 9-7 Ambrosia pollen at ±1000×. Scanning electron microscope photograph.

and *Xanthium.* The furrows are scarcely larger than the pores. *Ambrosia* and *Iva* are difficult to separate since their size, spines, and furrows are similar. Under oil immersion, the surface of *Iva* pollen may be more distinctly granular between the spines. *Ambrosia* pollen counts from samples near local sources of *Iva* should be made with caution. *Xanthium* pollen is of two distinct types. A few species, chief among them *X. spinosum,* are similar to the *Ambrosia-Iva* type. Other species of *Xanthium,* including the widespread *X. strumarium,* are distinguished by their larger size and smaller spines. The spines are often mere vestiges. The pollen diameter of these latter species is generally greater than 26 microns which should separate them from most species of *Ambrosia.* There is a slight size overlap with *A. psilostachya.*

Ambrosia pollen may rarely be much larger than normal. These grains usually have more than three furrows and may be from hybrids. Except perhaps in some western and coastal areas, recognition of *Ambrosia* pollen in North America presents no great difficulty.

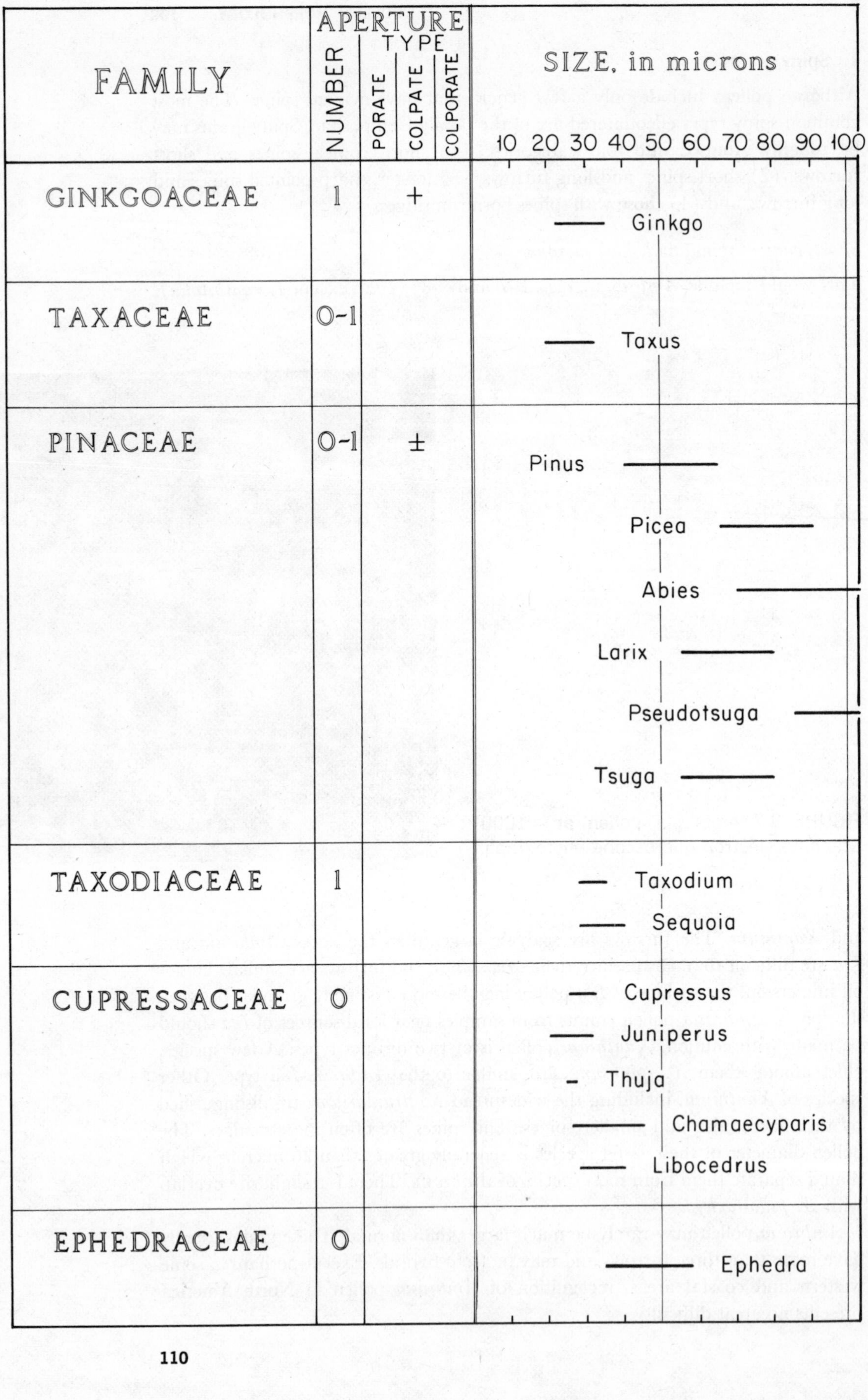

FAMILY	APERTURE NUMBER	TYPE: PORATE	TYPE: COLPATE	TYPE: COLPORATE	SIZE, in microns (10 20 30 40 50 60 70 80 90 100)
GINKGOACEAE	1		+		Ginkgo
TAXACEAE	0-1				Taxus
PINACEAE	0-1		±		Pinus Picea Abies Larix Pseudotsuga Tsuga
TAXODIACEAE	1				Taxodium Sequoia
CUPRESSACEAE	0				Cupressus Juniperus Thuja Chamaecyparis Libocedrus
EPHEDRACEAE	0				Ephedra

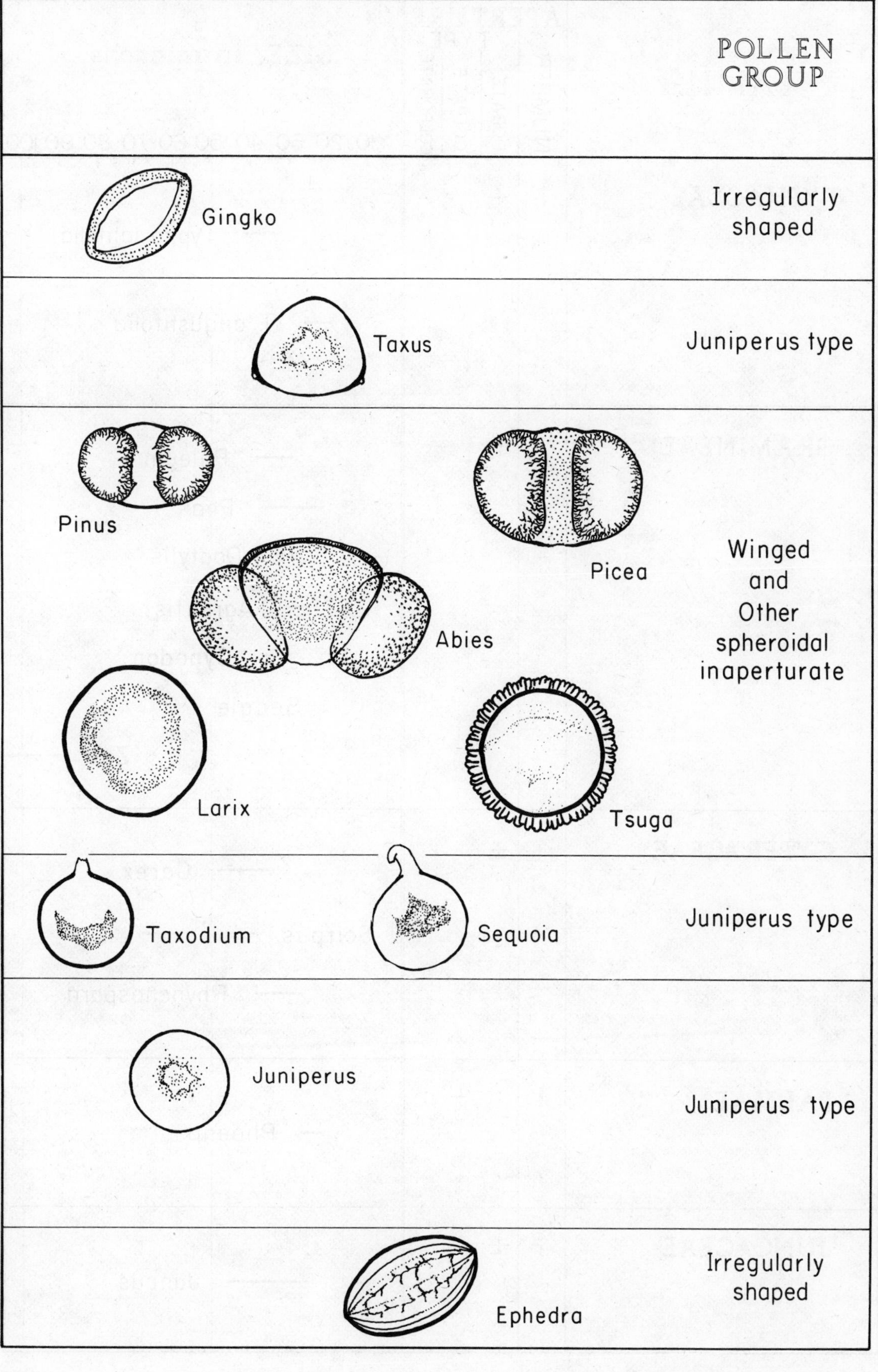
POLLEN GROUP
Gingko
Irregularly shaped
Taxus
Juniperus type
Pinus
Picea
Abies
Winged and Other spheroidal inaperturate
Larix
Tsuga
Taxodium
Sequoia
Juniperus type
Juniperus
Juniperus type
Ephedra
Irregularly shaped

FAMILY	APERTURE NUMBER	TYPE: PORATE	TYPE: COLPATE	TYPE: COLPORATE	SIZE, in microns (10 20 30 40 50 60 70 80 90 100)
TYPHACEAE	1	+			Typha latifolia; T. angustifolia
GRAMINEAE	1	+			Phleum; Poa; Dactylis; Agrostis; Cynodon; Secale; Zea
CYPERACEAE	?	±			Carex; Scirpus; Rhynchospora
PALMAE	1		+		Phoenix
JUNCACEAE	?	±			Juncus

	POLLEN GROUP
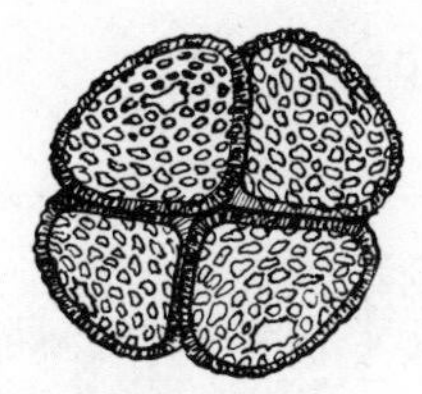 Typha latifolia 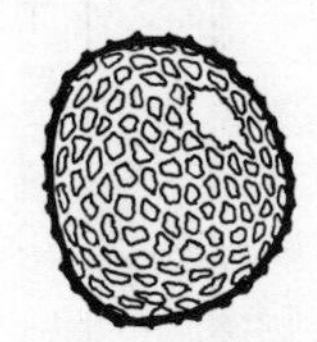 T. angustifolia	Irregularly shaped and netted
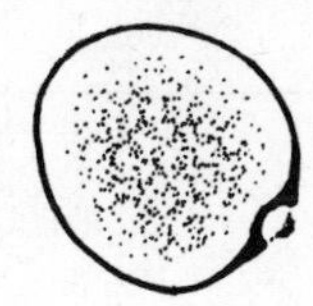 Phleum 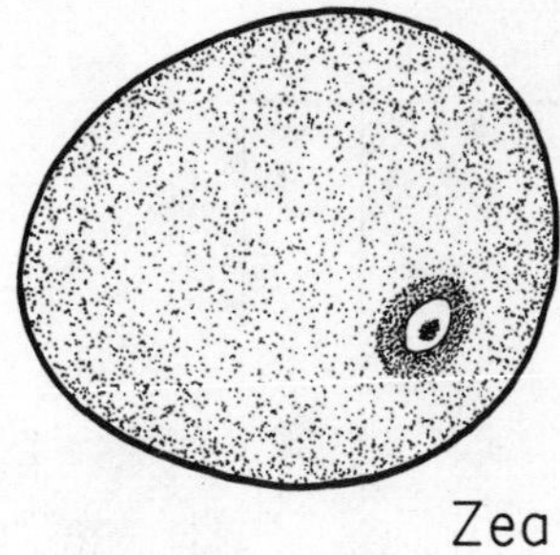 Zea 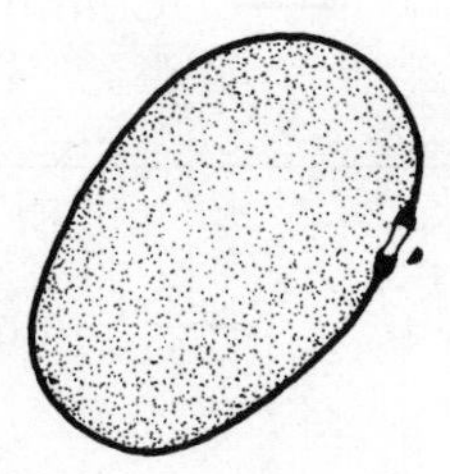 Secale	Porate
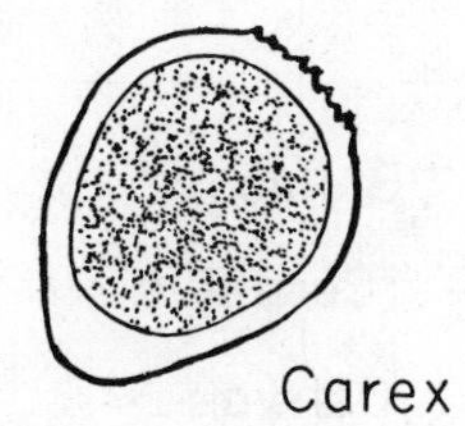 Carex Rhynchospora	Irregularly shaped
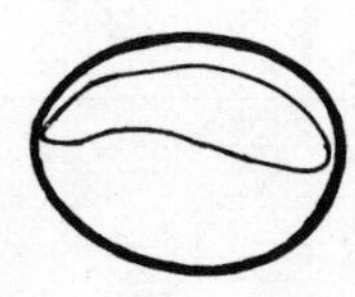 Phoenix	Irregularly shaped
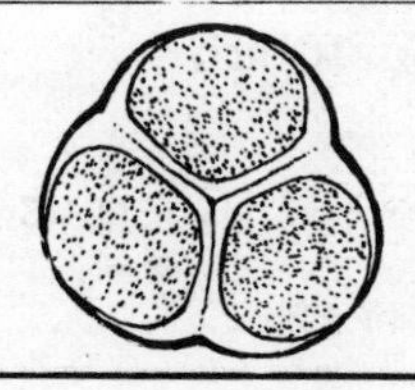 Juncus	Irregularly shaped

FAMILY	APERTURE NUMBER	TYPE PORATE	TYPE COLPATE	TYPE COLPORATE	SIZE, in microns (10 20 30 40 50 60 70 80 90 100)
CASUARINACEAE	3	+			Casuarina
SALICACEAE	0-3		+		Salix Populus
MYRICACEAE	3	+			Myrica Comptonia
JUGLANDACEAE	3-15	+			Juglans Carya
BETULACEAE	3-5	+			Betula Carpinus Ostrya Alnus Corylus
FAGACEAE	3		+	+	Fagus Castanea Quercus

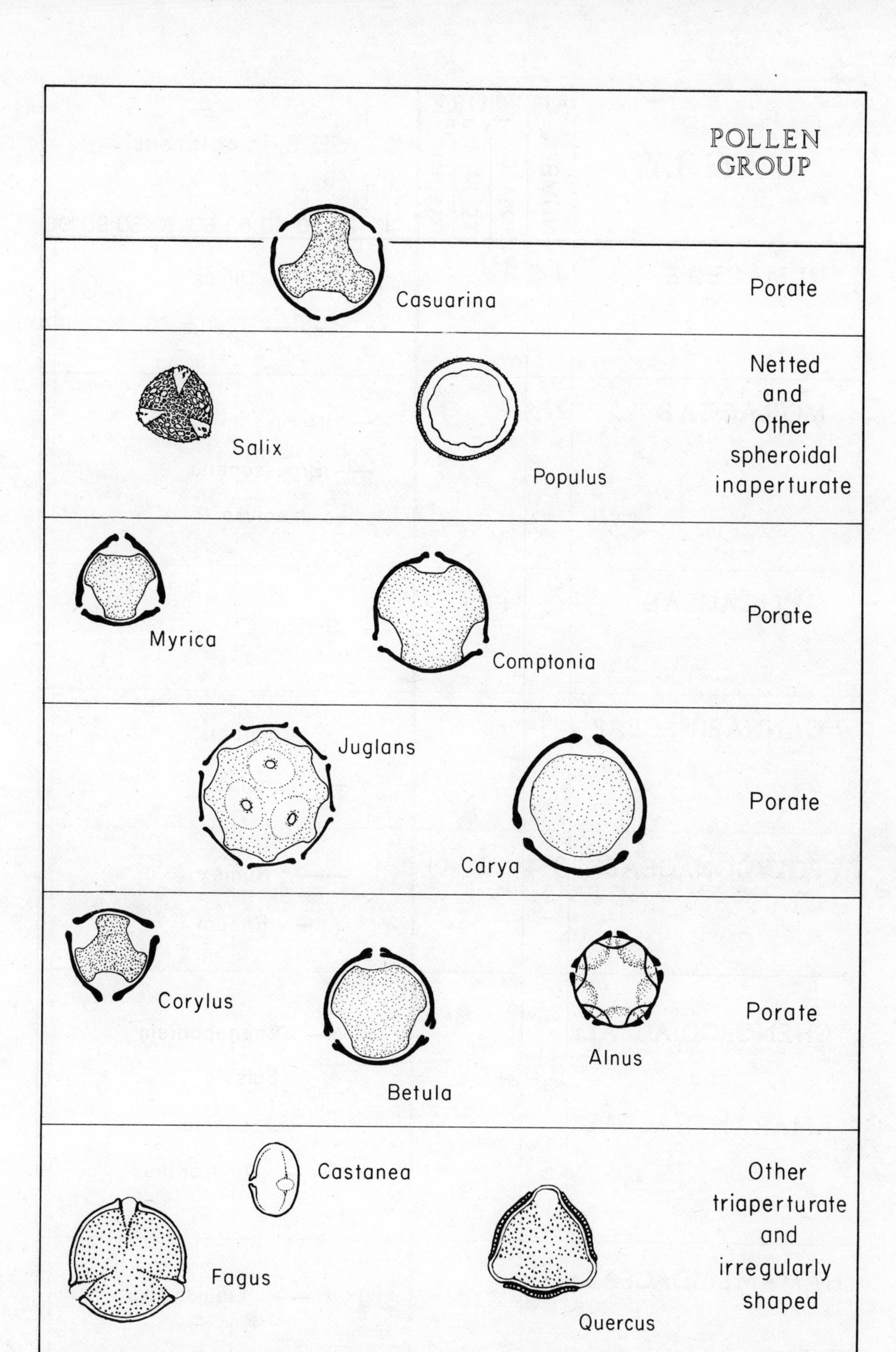
POLLEN GROUP
Casuarina
Porate
Salix
Populus
Netted and Other spheroidal inaperturate
Myrica
Comptonia
Porate
Juglans
Carya
Porate
Corylus
Betula
Alnus
Porate
Castanea
Fagus
Quercus
Other triaperturate and irregularly shaped

FAMILY	APERTURE NUMBER	APERTURE TYPE: PORATE	APERTURE TYPE: COLPATE	APERTURE TYPE: COLPORATE	SIZE, in microns (10 20 30 40 50 60 70 80 90 100)
ULMACEAE	4-5	+			Ulmus Celtis
MORACEAE	2-3	+			Morus Broussonetia Maclura
URTICACEAE	3	+			Urtica
CANNABINACEAE	3	+			Cannabis Humulus
POLYGONACEAE	3-4			+	Rumex Rheum
CHENOPODIACEAE and AMARANTHACEAE	8-90	+			Chenopodium Salsola Kochia Amaranthus
HAMAMELIDACEAE	12-20	+			Liquidambar

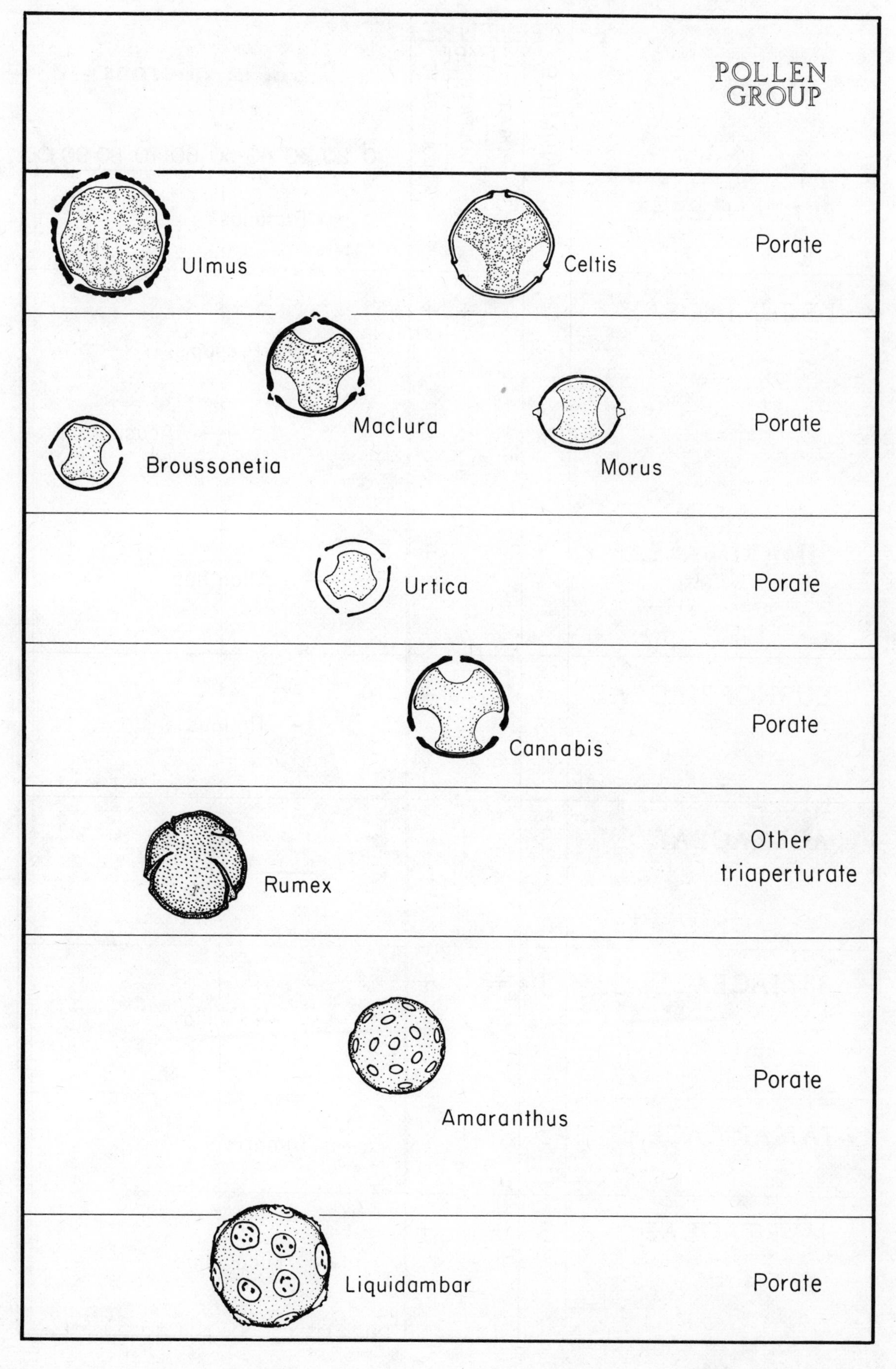
POLLEN GROUP
Ulmus
Celtis
Porate
Maclura
Broussonetia
Morus
Porate
Urtica
Porate
Cannabis
Porate
Rumex
Other triaperturate
Amaranthus
Porate
Liquidambar
Porate

FAMILY	APERTURE NUMBER	APERTURE TYPE: PORATE	APERTURE TYPE: COLPATE	APERTURE TYPE: COLPORATE	SIZE, in microns (10 20 30 40 50 60 70 80 90 100)
PLATANACEAE	3		+		Platanus
LEGUMINOSAE	3		+	+	Prosopis Acacia
SIMARUBACEAE	3			+	Ailanthus
EUPHORBIACEAE	3			+	Ricinus
ACERACEAE	3		+		Acer
TILIACEAE	3	+		+	Tilia
TAMARICACEAE	3		+		Tamarix
MYRTACEAE	3			+	Eucalyptus

	POLLEN GROUP
 Platanus	Netted
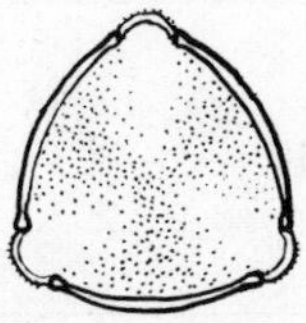 Prosopis 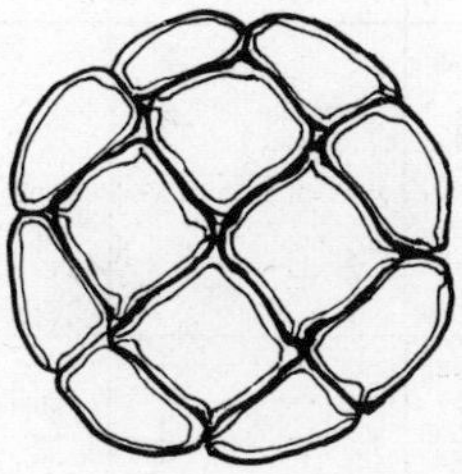 Acacia	Other triaperturate and irregularly shaped
 Ailanthus	Netted
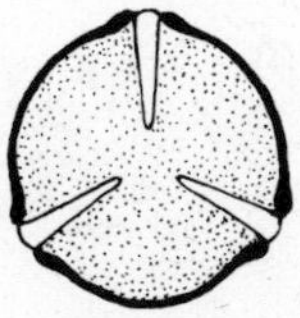 Ricinus	Other triaperturate
Acer 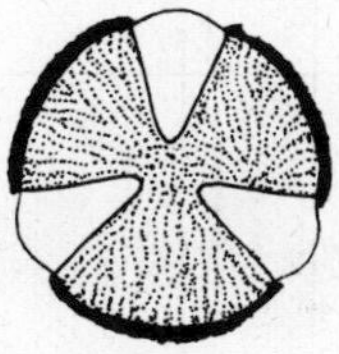	Other triaperturate
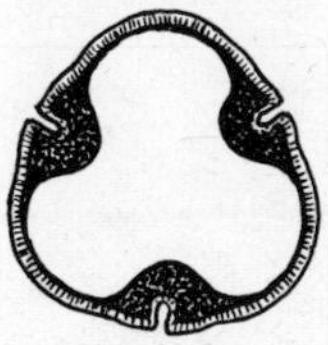 Tilia	Porate
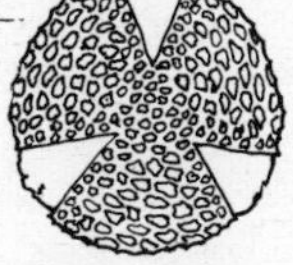 Tamarix	Netted
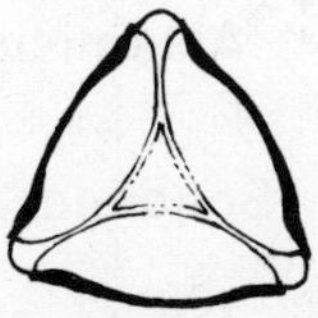 Eucalyptus	Other triaperturate

FAMILY	APERTURE NUMBER	PORATE	COLPATE	COLPORATE	SIZE, in microns (10 20 30 40 50 60 70 80 90 100)
UMBELLIFERAE	3			+	Daucus
NYSSACEAE	3			+	Nyssa
GARRYACEAE	3			+	Garrya
OLEACEAE	3-4		+	+	Fraxinus Ligustrum Olea
PLANTAGINACEAE	4-14	+			Plantago
COMPOSITAE	3			+	Ambrosia Iva Xanthium Artemisia Solidago

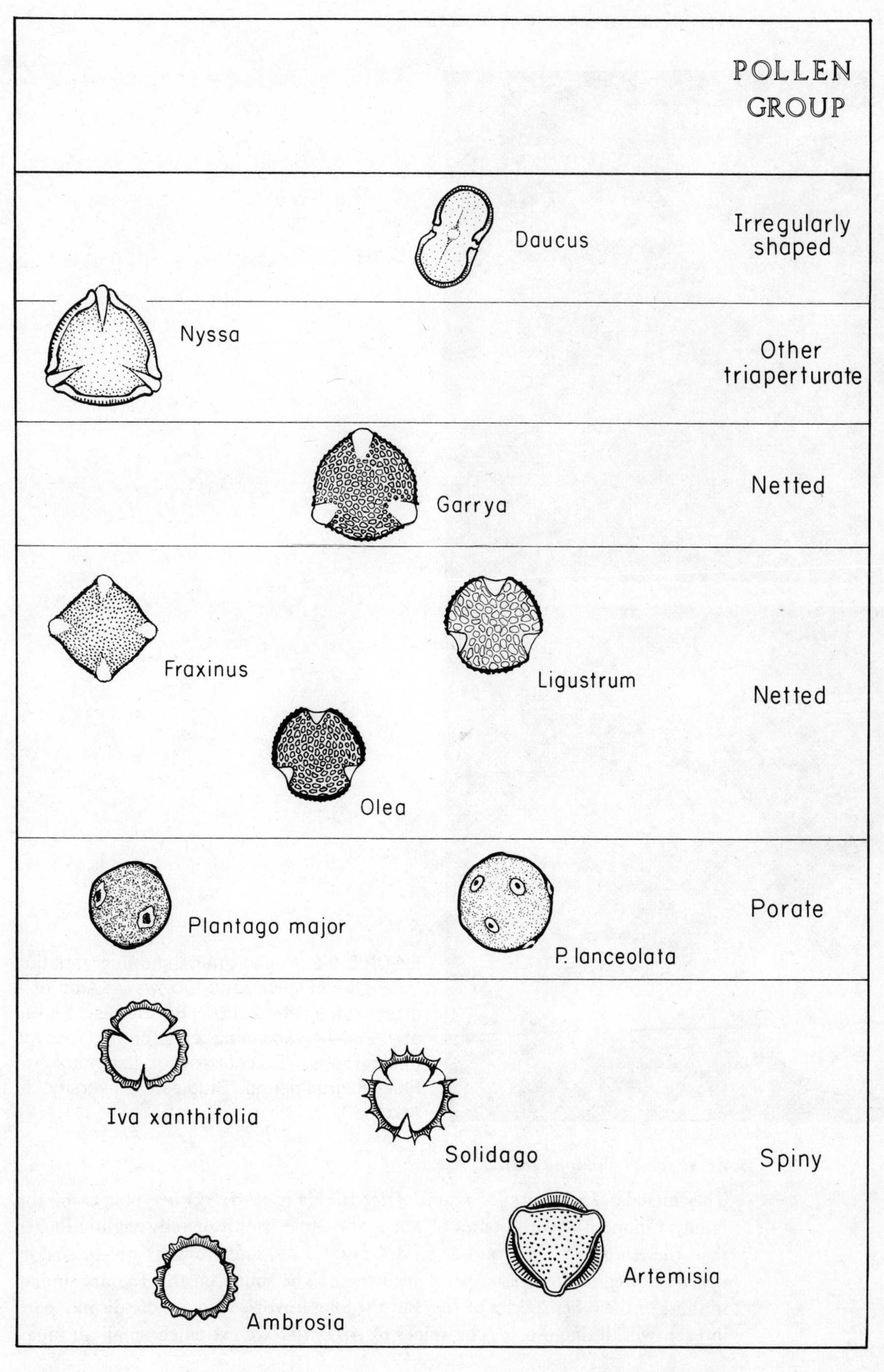
POLLEN GROUP
Daucus
Irregularly shaped
Nyssa
Other triaperturate
Garrya
Netted
Fraxinus
Ligustrum
Netted
Olea
Plantago major
P. lanceolata
Porate
Iva xanthifolia
Solidago
Spiny
Artemisia
Ambrosia

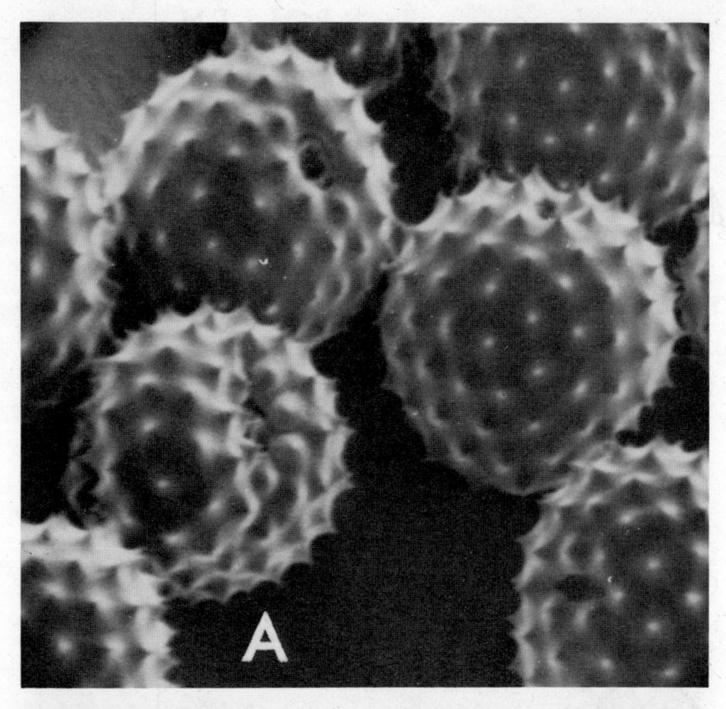

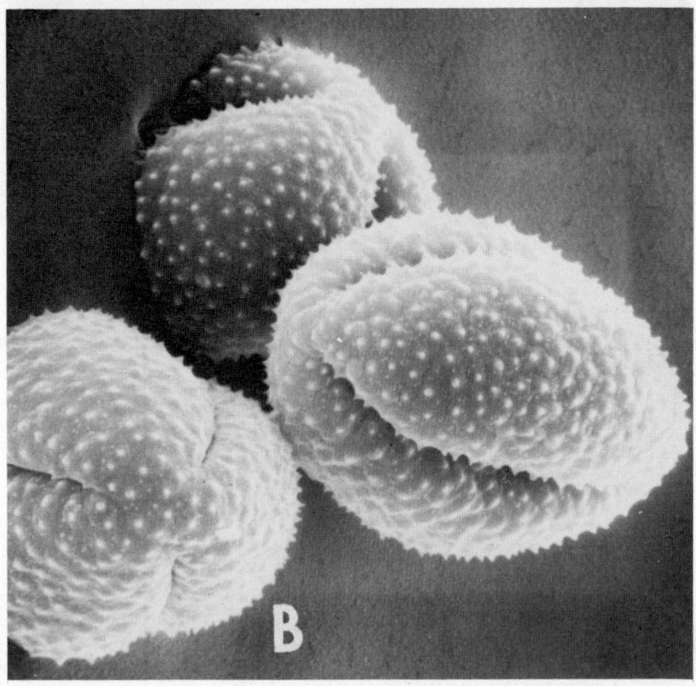

FIGURE 9-8 Pollen grains showing variation in lengths of spines and furrows. A, Ambrosia artemisiifolia at ±2200×. B, Artemisia tilessii at ±2500×. Scanning electron microscope photographs. (B, courtesy of Laboratory of Paleoenvironmental Studies, University of Arizona.)

Short spines and long furrows

This includes *Iva xanthifolia* and *Artemisia* (Fig. 9-8). Other pollens of the family Compositae with spines of about the same length or only slightly longer than the *Ambrosia-Iva* group (e.g., *Solidago, Aster,* and *Conyza*) are treated in the next category for reasons explained there. The spines on this *Iva* are similar to those of the other species of *Iva,* but the long furrows with a more distinct pore in each will distinguish it. The spines of *Artemisia* are extremely small and may appear as granules rather than spines. A distinguishing characteristic of *Artemisia* pollen is that, in optical cross section, the exine at the midpoint between furrows is appreciably thickened.

Longer, sharp-pointed spines and long furrows

These are from *Solidago, Aster, Conyza, Baccharis, Helianthus,* and others which are occasionally found in samples. These are not easily distinguished from each other, even though extremes of spine length may vary from over seven microns for some species of *Helianthus* to slightly more than one micron for *Gutierrezia.* Such short-spined grains as *Gutierrezia,* some species of *Solidago* and *Aster,* and others are considered in this rather than a short-spined category because of an *appearance* of greater spininess. This is caused by narrower spine bases or acuminate tapering giving the pollen a sharper-spined bristly look. Admittedly, this is a somewhat subjective character which may be modified by various effects of the mountant. If, for example, *Ambrosia* and *Iva* are overexpanded, the spines may appear to be acuminate. Knowledge of the local flora and proximity of sources to the sampler may help to resolve some problems of identification.

Spines on ridges

Among pollens with this character, *Taraxacum* is a common representative. These are strikingly ornate grains with spines borne on ridges and with large, smooth, nonstaining areas. The term fenestrate (windowed), applied to these grains by Faegri and Iversen (1964), is quite descriptive. Plants with this type of pollen are pollinated mostly by insects, but the pollen is occasionally airborne.

2. Winged grains

These are distinct, easily recognized pollen types. In the United States, most species of *Pinus, Picea,* and *Abies* have two wings as does *Tsuga mertensiana.* Other species of *Tsuga* are wingless in this sense. Some areas may have introduced plants with winged pollen (e.g., *Cedrus* and *Pseudolarix*), and the normally two-winged types may have aberrant individuals with one or several wings.

The exine of winged pollen grains is modified to form bladderlike lateral extensions of the broadly ellipsoidal body of the grain (Fig. 9.9). These are

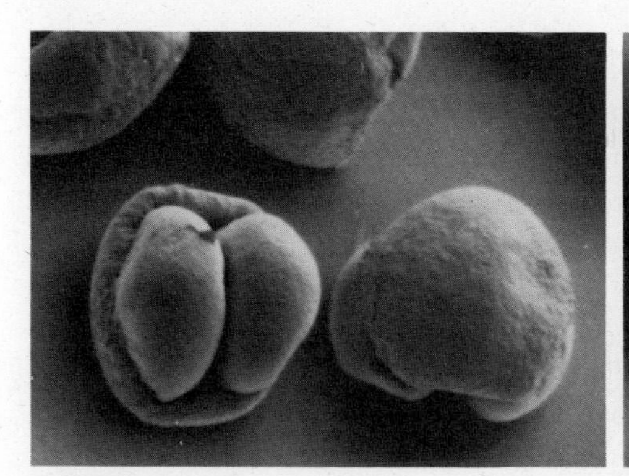
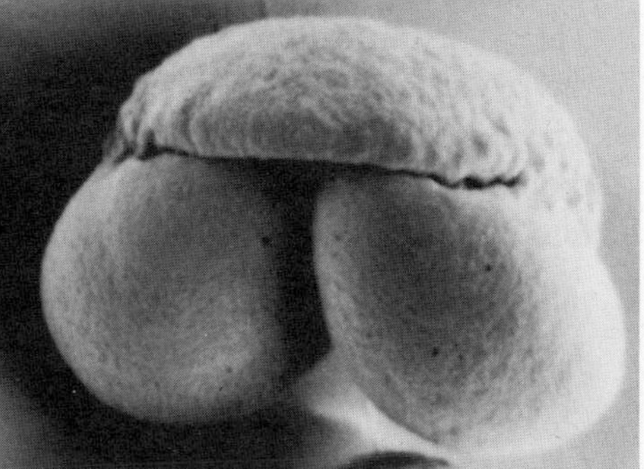

FIGURE 9-9 Pinus pollen. The two on the left are in polar view at ±500×. The one on the right is in equatorial view at ±1000×.

the wings which are also called bladders, sacci, vesicles, etc. They have an elaborate netted appearance. This is a subexinous feature and, consequently, their surface will appear rather smooth when seen in scanning electron micrographs. The more strongly ornamented and thicker-walled part of the body of the grain is the proximal portion, called cap or cappus. Directly opposite this is the thin, distal wall: the furrow area through which the germ tube emerges.

Characters used in separating genera (and perhaps groups of species) of winged pollen grains are: size, cap, wings, re-entrant angle, and furrow.

Size

Usually this refers to the long dimension of the body of the grain, but in some descriptions includes the wings or breadth of grains (Ting 1966). Size of *Pinus* (excluding wings) may range from 45 to 65 microns, of *Picea* from 68 to 91, of *Abies* from 78 to 111 (but mostly over 90), according to Wodehouse (1935). Ting records body sizes of *Pinus coulteri* and *P. jeffreyi* greater than 80 microns, and other authors have reported *Abies* pollen as large as 135 microns. Some of these larger sizes may be attributed to processing methods.

Cap

The cap will vary in fine sculpturing detail, but probably this is not a useful character for routine identification. Thickness of the cap wall is generally greater in typically larger grains, and the very thick (to 7 microns) and often irregularly thickened wall of *Abies* is a useful identification aid. The central portion of the cap of most species of *Abies* bears a faint triradiate streak. This is not found in other conifers nor in the balsam fir (*A. balsamea*) of eastern United States.

Wings

Pinus pollen has the largest wings relative to the size of the body, with *Abies* having the smallest, and *Picea* being in between. The wings of *Abies* are spheroidal, those of *Pinus* are spheroidal to somewhat ellipsoidal, *Picea* wings are spheroidal to ellipsoidal with a broad area of attachment to the body of the grain and often with a concavity in the surface opposite the cap. This may cause the wings to appear pointed in lateral (equatorial) view, but they may vary in fresh material depending on mounting techniques.

Re-entrant angle

The angle formed by the juncture of the wings and cap in lateral view is termed the re-entrant angle. This is sharp in *Pinus* and *Abies,* but appears as a continuous line or smooth curve in *Picea.* This juncture may be marked by a thickened projection of the cap (the marginal crest or ridge) in *Pinus.* Reduced crests are found uncommonly in *Abies,* but not in *Picea.*

Furrow

The furrow area of winged conifer pollen is the smooth, flexible, thin wall opposite the cap and between the distal roots (extremities) of the wings. A distinct furrow is seldom seen in *Pinus* or *Abies,* but is common in many species of *Picea.* There it appears as a V-shaped indentation on grains in lateral view or as two long, parallel lines on those in polar view, often with constrictions of the outline at their extremities. While the wall at the furrow area of all of these pollens is relatively smooth, the presence of slight excrescences has been used to separate the subgenus Haploxylon from other *Pinus* in fossil material. Apparently, this condition has not been exploited in fresh material.

These distinctions between common winged pollens have primarily concerned grains oriented in lateral (equatorial) view. Most winged pollen trapped on microscope slides or other plane surfaces comes to rest on the cap. Without turning these would be in polar view to the microscopist, with the furrow area uppermost. Pollen of *Pinus* and *Abies* in this view usually appears as three distinct globular bodies with the body of the grain sandwiched between the two wings. This is less noticeable in *Picea* with these grains presenting a more-or-less ellipsoidal outline. For pollen trapped in aerobiological studies, *Pinus* will be found frequently in either orientation while *Picea* and *Abies* are more commonly seen in polar view. The orientation probably depends to some extent on the type of adhesive, the type of mounting medium, and the thickness of mounts.

3. Juniperus-type grains

Plants producing this type of pollen include a number of native and introduced evergreens. The genus *Juniperus* is the most widespread of these in the United States.

The pollens of these plants are spherical with a thin exine which stains lightly (often lavender with basic fuchsin). They have thick, irregular intines whose contact with the cytoplasm is usually angular, varying from a star-shaped to a rectangular outline. This is the outstanding identifying feature. The surface bears small granules which stain more deeply than the exine. These vary in number and are said to be deciduous in some species. When wetted, these grains frequently expand greatly causing the exine to rupture. The entire outer wall is often cast off.

Pollens of the Cupressaceae are mostly unmodified from this form, although some may have faint pores, probably of no diagnostic importance in fresh material. Pollens of the Taxaceae and Taxodiaceae are modified to show areas of varying distinctiveness. In *Taxus,* there are two minute beadlike thickenings of the exine. With optimum staining, these show a faint arching connection, probably marking a dehiscence zone. These grains are frequently somewhat angular with a flattened side and an opposite rounded side corresponding to the curve of the arching line. This arching line may be seen in only a few of the pollens of ornamental yews (see illustration for *Taxus baccata* in Hyde and Adams 1958: 104), but it is common in *Taxus americana.* Because of this inconsistency, the feature probably is useful only in determining the presence or absence of *Taxus* pollen in a given field-collected sample and could not be used in quantitative assessments. This modified zone is considerably more pronounced in slightly immature pollen dissected from anthers. *Torreya,* of limited distribution in Florida and California, has a broad, often bulging, area termed the leptoma (Erdtman 1965).

The Taxodiaceae have papillae varying in size from slight bulges in *Taxodium* to fingerlike projections in *Sequoia.* Pollens with papillae are found in related ornamentals such as *Cryptomeria* and *Metasequoia.*

Unmodified, spherical, *Juniperus*-type pollen (members of the Cupressaceae) might be separated into broad groups by time of shedding and pollen size. Angular grains of this general type should be examined at high magnification to determine the presence or absence of *Taxus* in the sample. The genera of the Taxodiaceae may be identified by the form of the papillae.

4. Other spheroidal, inaperturate grains

Except for *Juniperus*-type grains, few pollens without apertures are encountered. Some of those discussed here possess subtle features which are considered analogous to more fully developed apertures. Some grains, mostly porate, have indistinct apertures and might be thought of as inaperturate by the inexperienced.

Large pollen grains, mostly 65 microns or greater, are found in *Larix, Pseudotsuga,* and *Tsuga* of the Pinaceae. The exines of *Larix* and *Pseudotsuga* may rupture on mounting in aqueous media. They are smooth and stain magenta in basic fuchsin. The intine is very thick, but relatively less so than in the *Juniperus*-type pollen. The outline of the interface between the intine and the cytoplasm may be angular, but is generally rounded. The interior of the grains appears granular or crowded with spherical bodies. Perhaps this varies with the age and condition of the pollen. The pollen of *Pseudotsuga* is generally larger (90–100 microns) than that of *Larix,* but local and regional collections should be studied before this character is used for identification.

Other pollens of this general type are occasionally encountered in local situations. For example, pollen of the monkey puzzle tree, a South American conifer of the genus *Araucaria,* might be expected in warm regions where it is planted as an ornamental.

Tsuga (except *T. mertensiana,* mentioned under winged grains) appears as a coarsely roughened, spheroidal grain in field-collected samples or reference microscope slides. One surface is thicker and more highly sculptured than the other. The exine is convoluted into irregular islands, appearing as coarse rods in optical cross section. Very fresh pollen may not be wetted by the mounting medium and will fail to stain. Such grains will be yellowish-brown. In thick mounts, grains may be turned ninety degrees from the normal view, in which case they appear as two hemispheres with the smaller, less sculptured one fitted into the other. Pollen morphologists consider the thicker-walled portion to be analogous to the cap of winged pollen grains with its margin delimited by a "puffy frill," the equivalent of the wings. With this interpretation, the thinner-walled hemisphere would serve the function of the furrow area of winged grains. In out-of-season and presumably refloated pollen, the grains are frequently broken with the smaller hemisphere lost.

Populus pollen is spheroidal, inaperturate, and relatively featureless, probably as a result of reduction of form in response to adaptation from primitive insect pollination to present day pollination by wind. Its close relative, *Salix* (Fig. 9-10), is elaborately sculptured, fairly thick-walled, and effectively insect-pollinated. Because of its thin walls, *Populus* pollen tends to rupture easily and, at times, suggests a more complex structure. The thin, granular wall stains lightly and may break into irregular plates showing the unstained intine beneath. The intine and cell contents may be withdrawn from the exine forming a space that gives the

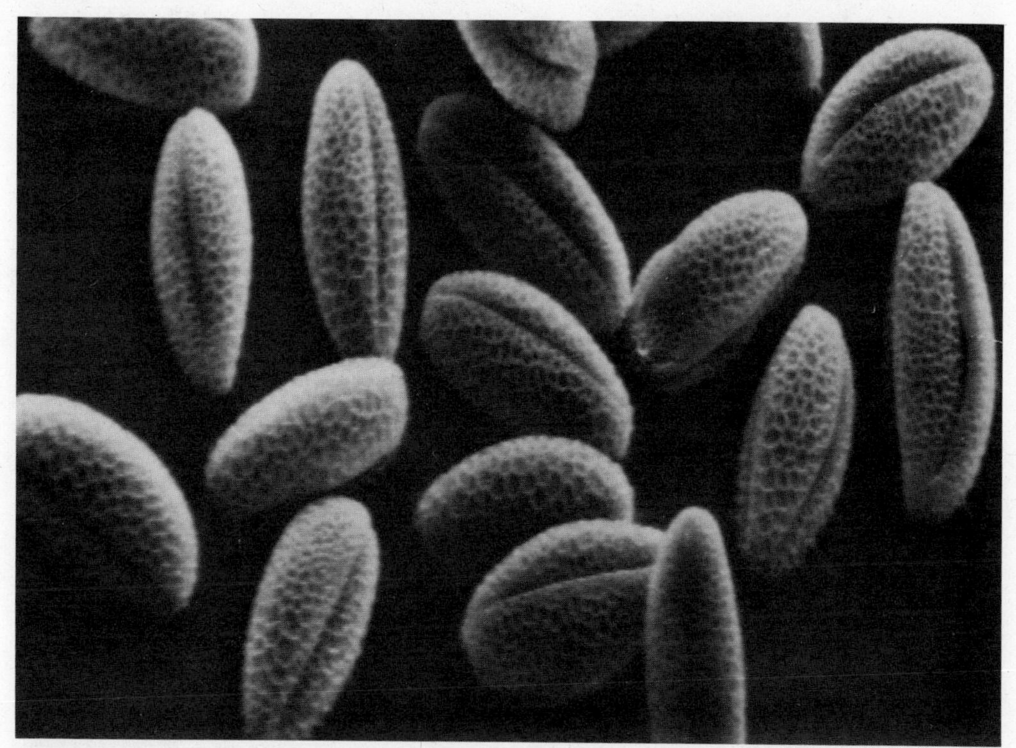

FIGURE 9-10 Salix pollen at ±1400×. Scanning electron microscope photograph. Courtesy of Brookhaven National Laboratory.

appearance of a thick wall in optical cross section. Ruptures or wrinkles of the outer wall may imitate furrows. Old pollen, especially that caught on rotating impactors, may disintegrate almost completely showing little typical pollen form.

Abortive pollens of many plants may be smaller than normal and lack apertures. Sometimes fairly large quantities of abortive *Quercus* pollen, perhaps from sterile hybrids, is collected.

Grass pollen has a characteristic pore, but the orientation may be such that the pore cannot be seen. Abortive grasses may occasionally be smooth spheres without a trace of a pore.

5. Porate grains

Over 40 percent of the genera listed in Chapter 3 are included here. These are pollens whose outstanding characteristic is their possession of pores. There are not many distinct types since many of the genera are found in a few families. These include the grasses (Gramineae), a number of trees (chiefly Betulaceae, Juglandaceae, and Ulmaceae), and weeds such as the pigweeds (Chenopodiaceae and Amaranthaceae). A few other porate grains have some characteristics more outstanding than the form of their apertures and these are considered elsewhere.

Distinctly porate grains may be classified by the number, position, form, and size of the pores and associated features such as thickened walls or the presence of opercula.

The grasses have a single pore surrounded by a thickened ring of exine. In expanded grains, the colorless pore membrane protrudes, bubblelike, bearing a usually circular, thickened lid. This lid is the operculum, which accepts stain in the same manner as the exine. These features are best seen in grains with a pore in optical cross section. There, the thickened pore rim is easily detected and the operculum is seen as a hemispherical body floating on the pore membrane. The operculum is frequently lost in aqueous mounts and is sometimes seen as a half-raised lid as though hinged (Fig. 9-4). The morphology of grass pollen is monotonously uniform. They range in size from 22 to over 100 microns, and their shape may vary from spheroidal to ovoid. Most of the grass pollens reported as important in hayfever range in size from about 25 to a little over 40 microns. Johnson grass (*Sorgum halapense*) is from 40 to 55 microns. Grass pollen is described as smooth, but is actually minutely granular or reticulate. Positive identification of these grains is not possible except in isolated instances. *Secale cereale* may be identified (at least in some regions) by its large size, ellipsoidal shape, position of the pore toward one end of the ellipse, and flowering period. Unlike many cereal pollens, it is well dispersed. Fossil pollen analysts have separated some grass groups chiefly by phase contrast microscopy which shows fine-sculpture detail. The special key for the Gramineae in Faegri and Iversen (1964) should be studied to determine its adaptability to aeropalynological problems. There are very many grass species, but only a few occur in large enough quantities or produce enough pollen to significantly contribute to total grass pollen in the atmosphere. A knowledge of the abundance of these species, their flowering period, and their pollen size range will give the investigator an idea of the primary producers in a locality at certain times. Cereal grass pollens are usually appreciably larger than wild types, being generally greater than 50 microns. *Zea* (corn or maize) pollen is the largest, often more than 100 microns. It might be expected that atmospheric grass pollen counts would be high in grain-growing areas, but these cultivated forms are often self-pollinated, or produce little pollen, or are dispersed over short distances. Agricultural practices may complicate the interpretation of grass counts by refloating pollen during harvesting.

Other monoporate pollens such as *Typha* and *Juncus* have indistinct pores or other identifying features and are discussed later.

All of the two- to five-pored grains discussed here, with the exception of *Carya, Ulmus,* and *Planera,* are aspidate. The aspides are shield-shaped areas surrounding the germ pores. These are a result of thickenings of the intine beneath the pore and, often, lesser thickenings of the exine in this region. Usually these areas appear smooth and contrast in color (in stained material) from the frequently granular-appearing cell contents. The families Casuarinaceae, Myricaceae, Betulaceae, Moraceae, Urticaceae, *Juglans* of the Juglandaceae, and *Celtis* of the Ulmaceae all have aspidate pollen grains. There are isolated occurrences of this type of pollen in other dicotyledonous families. As might be expected, characters of the aspides themselves are not of much value in pollen identification, but they are easily recognized and many three-pored, aspidate grains are broadly categorized as "betuloid" in studies of airborne pollen. However, there are many characters by which aspidate grains can be more precisely classified. A few are two-pored. These include *Morus* and *Broussonetia* of the Moraceae, occasionally *Urtica* and, rarely, aberrant grains of *Betula. Morus* is typically two-pored (occasionally three- or four-pored), broadly elliptical or spheroidal, and less than 20 microns in longest dimension. The colorless pore membrane often protrudes cylinderlike and is capped by a delicate operculum. The pores are situated at the ends of the long axis of the broad ellipse and are not exactly opposite each other. *Broussonetia* is much like *Morus,* but is smaller, being about 13 or 14 microns in long dimension. It is probably predominantly two-pored, but is more commonly three- or four-pored than is *Morus. Urtica* is also similar, but is commonly three-pored and 10 to 13 microns in greatest dimension. The thickenings of intine beneath the pores are less pronounced than in *Morus* and *Broussonetia.* Other members of the Urticaceae may be components of the air-spora in local situations. These and all members of the Moraceae, Urticaceae, and Cannabinaceae have thin exines, at least relative to other aspidate pollens considered here. They stain less intensely than do grains of the other families having aspidate pollen. The Moraceae and Urticaceae usually appear lavender to pale red when stained with basic fuchsin. *Maclura* is a little larger than other members of its family. It is typically three-pored with a considerable number of four-pored grains. Some collections are predominantly four-pored. The Cannabinaceae stain pale red to red in basic fuchsin, flower in summer rather than spring, and are larger (about 25 microns) than the Moraceae and Urticaceae. *Morus* pollen may be trapped when other, more conspicuous pollen types are in the air in quantity and are likely to be obscured or otherwise overlooked. The delicate *Urtica* grains collapse easily and may not be recognized as pollen if grains are refloated or if there is much delay in slide preparation.

The considerable number of generally three- or four-pored, brightly stained, aspidate grains can be separated to some degree by considerations of flowering time, thickenings of the exine at the pore, shape in polar view, and size. However, it is not possible to provide precise criteria for their identification, and characteristics for all aspidate grains found in any one region should be tabulated and studied. Close attention should be paid to the form of the exine at the pore. In *Betula,* for example, there is a knoblike thickening of the exine. This and cognizance of wall stratification give it an appearance in cross section that has been compared to "opposing snake heads." In other members of the Betulaceae (except *Alnus*), there is a reduction in the amount of this thickening, the pores protrude less, and the outline in polar view becomes more rounded. *Corylus,* however, with less thickening than *Betula,* frequently appears triangular. Their shape has been used for identification in fossil material, but appears to vary with mounting procedures in fresh pollen and, by itself, is not a reliable characteristic. *Myrica* and *Comptonia* of the Myricaceae are also commnoly triangular but with a thinner exine at the pore. *Comptonia* and also *Carpinus* and *Ostrya* of the Betulaceae are usually 27 microns or greater (rare giants of *Comptonia* may exceed 40 microns), often more than three-pored, and with little modification of the exine at the pore. There is a tendency for *Carpinus* and *Ostrya* to be more round in polar view than other members of the Betulaceae and *Comptonia;* however, variation precludes this as an identifying characteristic. *Alnus* is easily identified. The pores protrude more than those of other grains giving the usually four- or five-pored grains (sometimes three- or six-pored) a polygonal appearance in polar view. These grains possess curving bands of thickened exine (arci) connecting adjacent pores. When properly stained, they are very distinctive. Arci may sometimes be seen in greatly overexpanded mounts of other pollen grains of the Betulaceae (e.g., *Corylus californica*). *Casuarina* pollen is much like that of *Corylus.*

Celtis is mostly larger than other aspidate grains mentioned (usually greater than 30 microns). They have a distinctive granular appearance and the pores do not protrude. Wodehouse (1935) states that *C. occidentalis* is highly variable in size and pore number with large numbers of abortive grains, all suggestive of hybrid origin. Normal grains might be confused with those of *Comptonia, Carpinus,* and *Ostrya.*

These features of aspidate grains are best seen in polar view. Fortunately, this is their usual orientation in air samples. The characters of shape and exine thickening, particularly, are less helpful for these pollens in equatorial view and identification of such grains is difficult.

Juglans is the remaining aspidate pollen to be considered. These grains have 6 to 15 pores. The pores are mostly in one hemisphere. Those situated around the perimeter give the grains an angular appearance in polar view. They are easily identified to genus and in some regions may be identified to species by pore number or pollen size. They are usually 35 microns or more in greatest dimension.

The closely related *Carya* has three (but occasionally four or more) nonaspidate pores. The pollen is large (about 40–50 microns), of smooth texture, and spheroidal in polar view. The pores are arranged in a zone not quite in the plane of the equator. As with *Juglans,* some species may be identified by pore number or grain size.

Ulmus pollen grains are normally five-pored, but four-pored grains are common, and three-, six-, or seven-pored ones may be seen occasionally. The pores are not as prominent as in most of the preceding pollens, probably because of other distracting features. The grains are polygonal as seen in the usual polar orientation. The intine is thick, especially beneath the pores. The exine is fairly thick, deeply staining, and characteristically wavy. The degree of waviness may vary between collections. In field-collected samples, there may be considerable variation in pollen size, but they are mostly 30–35 microns.

Planera of the same family has similar grains. According to Wodehouse (1935), they can be distinguished by a usual complement of four rather than five pores and the presence of arci somewhat similar to those of *Alnus.*

There are a number of multipored grains common in air samples in which the pores are more-or-less scattered over the surface of the grains and are not restricted to the equator or other zones as are those of pollens previously mentioned. These are found in the families Chenopodiaceae and Amaranthaceae, *Liquidambar* of the Hamamelidaceae, and the Plantaginaceae. The pollens of the Chenopodiaceae and Amaranthaceae are of uniform morphology and are not routinely separated in examinations of air samples. They are sometimes listed under the convenience heading of Cheno-Ams. Pore numbers of various members from around the world vary from 8 to 90 and pollen size varies from 12 to 40 microns (Erdtman 1952). Most of the types trapped are intermediate in these values. Their general appearance has been described as "golf ball-like" (Solomon et al. 1967). A few of these types might be separated by their large pores (*Salsola*) or by the wide spacing of their pores (*Sarcobatus*). There is also variation in texture of the surface of the grains between the pores, with some appearing rather smooth while others are decidedly granular. Wodehouse (1935) mentions that granules on the surface of pore membranes may be fused to resemble an operculum in some species, and Hyde and Adams (1958) give measurements for the operculum of *Chenopodium album.* This is not an operculum in the sense described for the grasses. Heusser (1971) describes and shows photographs of pronounced reticulate forms for two species of South American Amaranthaceae.

Liquidambar is easily distinguished by having 12 to 20 large pores which vary in size and may be up to 10 microns in diameter. The colorless pore membrane is covered with granules which accept stain. The contrast is striking. The pore membrane in cross section at the perimeter of the grain appears as a coarsely roughened dome standing out from the exine as much as 2.5 microns. Grains may range in size from 28 to 38 microns in the same collection, averaging about 35 microns. Some of the smaller grains are obviously abortive, but others appear to be functional. The exine is pitted and stains a rich magenta in basic fuchsin, although it may require some time for freshly collected pollen to fully stain. Grains are spheroidal to somewhat angular, depending on the degree of expansion.

The Plantaginaceae are represented by a number of species of *Plantago*. Frequently, these can be determined to species, but the preponderance of pollen of this type in air samples are from *P. lanceolata,* with occasional small amounts from *P. major*. The most outstanding feature of *P. lanceolata* is the appearance of its pores in well-stained material. These are round, surrounded by a thickened rim, and with a small operculum centered on the pore membrane. Their appearance is "doughnutlike." Pores in cross section around the perimeter appear as small, colorless bubbles. In cleared material used for fossil pollen comparisons, the granular or minutely spiny character of the surface has been used as a key character in separating species, but this feature is not apparent in fresh material. Fresh pollen is usually described as smooth or of mottled appearance. The granular nature of the exine is distinct in overexpanded mounts. There is wide variation in size of *P. lanceolata,* but they are generally 25 to 30 microns. Dwarf, abortive grains are frequently found, causing problems for inexperienced pollen workers. The pollen of *P. major* is smaller (about 20 microns) with pores of different form and number. Its pores are fewer (four to six) and of irregular shape, often appearing as patches torn from the exine.

Tilia pollen appears porate to most observers and is grouped with porate grains here even though such a classification is not strictly accurate. The three apertures are compound structures, each composed of a short, eliptical furrow underlaid by a "pore." Grains are usually oriented in polar view and in optical cross section it is seen that the wall is much thickened at the aperture. The thickened area is characteristically darker in either stained or unstained material. The outline of the grains may be round or, more commonly, subtriangular. The apertures are located on the sides rather than on the angles of the triangle. Grains may be seen in equatorial view in thick mounts. In this view the elliptical furrow is obvious and the pollen is seen to be much flattened at the poles giving an elliptical outline. The average maximum dimension is about 35 microns for *T. americana*. Ornamental species may be smaller and flower a week or two earlier. In routine examinations, the surface would be considered smooth. It is actually minutely pitted and bears some resemblance to some of the more finely patterned grains in the netted group.

6. Netted grains

All grains showing a definite netlike pattern in which this character is the most imposing are included here. There are many technical differences in netted grains as to the nature of the elements composing the net, the size and shape of the meshes, and whether the pattern is a surface or subsurface feature. The term reticulate (and its derivatives) is more strictly defined by palynologists. The texts of Erdtman (1952) and Faegri and Iversen (1964), especially, should be consulted for a better understanding of this sculpture.

Typha has grains occurring singly or united in tetrads. The latter type is discussed with irregularly shaped grains. *Typha angustifolia* has single pollen. These have a delicate reticulum and the netlike analogy is overly descriptive, but the sculpture is rather noticeable, and it appears that mesh size varies in different collections. Grains are spheroidal, but many are somewhat angular. Size is variable, but most are in the 20- to 27-micron range. The single pore is conspicuous and of variable size and shape but averages about five microns. There is no distinct pore margin. The intine underlying the pore is thickened.

The remaining netted grains have three or four furrows, sometimes with pores or suggestions of pores.

The exine of *Platanus* has been described as finely pitted, but mesh size is variable and some appear netlike. The grains are small, about 20 microns, with occasional giants and dwarfs. When in polar view and expanded, they are easily identified. The furrows are short and broad, consequently there is a larger polar area. Distinctive granules which accept basic fuchsin stain cover the furrow membrane. There may be many *Platanus* grains which do not fully expand, and these can be difficult to distinguish.

Fraxinus pollen is mostly 20 to 25 microns. It has a fine-netted appearance, more pronounced than *Platanus,* with a little variation in mesh size in the several species found in the United States. *Fraxinus americana* is the most widely distributed. There are usually four short furrows flecked with granules (but not as prominently as in *Platanus*). Occasionally, grains are found with three, five, or more furrows. The furrows have no definite margin and present a jagged appearance. In polar view, the pollen is polygonal with the apertures at the angles. In its normal, expanded condition, there is little chance for confusion with other commonly airborne pollens, but unexpanded grains may bear a superficial resemblance to *Platanus.* The exine is thin, and older mounts may greatly overexpand causing some problems in identification. The mesh of such grains may be misinterpreted as granules by inexperienced workers and listed with such diverse types as *Quercus. Olea* and *Ligustrum* are other members of this family (Oleaceae) with airborne pollen, although they are chiefly insect-pollinated. As the trend toward entomophily progresses, the grains show more pronounced sculpture. *Ligustrum* has a larger and more elaborate mesh than *Olea,* but both are more highly sculptured than other members of this group. Unlike *Fraxinus,* these genera possess a pore centered in each furrow. The pores are distinct only in grains in equatorial view and are more easily seen in *Olea* than *Ligustrum.*

Salix has three-furrowed, spheroidal pollen tending to be ellipsoial in equatorial view if not fully expanded (Fig. 9-10). The genus is widespread with many localities supporting several species. The various species range in pollen size from about 14 to 24 microns with giants and dwarfs not uncommon, but with most grains ranging from 15 to 20 microns. Grains in polar view are usually spheroidal, but sometimes, when not expanded, are strongly three-lobed. The ellipsoidal form of the unexpanded pollen may have a longer dimension (polar axis) than that described. The furrows are relatively longer than those of *Platanus* and *Fraxinus,* and the meshes of the net are larger and more distinct. The angular meshes tend to decrease in size toward the furrows. *Tamarix,* found in arid and saline situations, has pollen of about 15 to 18 microns, which is similar to *Salix* in most respects.

Garrya has three fairly short furrows with a bulging pore in each. The limits of both furrow and pore are indistinct. The grains are about 30 microns, subtriangular in polar view and spheroidal, but somewhat flattened at the poles in equatorial view. Mesh size of the net is variable, but is larger than *Fraxinus* and smaller than *Salix.*

Ailanthus has three long furrows, each with a distinct round germ pore. The meshes of the net are smaller toward the furrows and the poles. They tend to be elongate and oriented in parallel rows poleward, giving some grains a vague striate appearance. The network is intermediate in distinctiveness, about like that of *Garrya.* Expanded grains are spheroidal or flattened at the poles and generally 25 to 30 microns in diameter.

7. Other triaperturate grains

These include pollens with furrows, or with furrows and pores, whose outer walls are generally not conspicuously ornamented. A large number of grains not discussed here, that are found infrequently in air samples, would fall within this broad category. In routine analysis, grains of this general type may be roughly categorized as smooth or granular. These differences in sculpture may not be immediately apparent to inexperienced workers.

Essentially smooth grains would include *Rumex, Rheum, Prosopis, Ricinus,* and *Eucalyptus* of those mentioned in Chapter 3. The pollens of *Rumex* and *Rheum* of the Polygonaceae are morphologically similar and are described as having pitted exines. The walls are thin, staining lavender in basic fuchsin. Furrows are usually three or four, long and slitlike with a small, ellipsoidal pore in the center of each. The most outstanding feature of these grains is the presence of tightly packed, globular starch bodies. It sometimes requires optimum staining to see the furrows and pores. *Rheum* and various species of *Rumex* can be separated to some extent by size and the number of apertures. The widespread *Rumex acetosella* usually has four apertures and is from 20 to 25 microns. *Rumex acetosa* is usually three-furrowed and is about 20 microns. Other species of *Rumex* are mostly three-furrowed with several greater than 25 microns. *Rheum* has three furrows and is about 30 microns. Grains are spheroidal or, often in the case of the four-furrowed *Rumex acetosella,* squarish or lozenge-shaped.

Prosopis is commonly subtriangular in polar view and spheroidal or flattened at the poles in equatorial view. There are three long and indistinct furrows resulting in a small polar area; in fact, the three furrows sometimes merge at the poles. Each furrow has an elliptical pore up to seven microns in long dimension. Wall thickenings around the pores give them a distinctly marginate appearance in properly oriented grains. The pore membrane protrudes and is roughened. Size varies but is mostly 27 to 32 microns in polar view, excluding the bulging pore membrane. The surface texture is finely granular but could be considered smooth.

Ricinus pollen is uniform in size and shape. They are from about 26 to 30 microns and spheroidal regardless of orientation. The three furrows are distinct, long and tapering and rather narrow with an elliptical germ pore in each. The pore is not sharply defined and is five to six microns in long dimension. The exine at the equator is thickened in a band which is about the width of the long axis of the pores. The cross section of the exine at the furrow in polar view appears like lips. The localized thickening of the exine apparently confers rigidity which prevents much distortion in overexpanded grains. The surface is finely pitted, but for practical purposes may be considered smooth in routine analyses.

Eucalyptus is considered an unimportant contributor to the air-spora, but it is occasionally reported. The grains are small (about 20 microns), smooth, with extensions of the furrows merging to form a characteristic triangular area at the poles. The exine is thickened in the region of the round to ellipsoidal pore and grains in polar view appear triangular. In equatorial view, the pollen is ellipsoidal with the ratio of the equatorial to the polar axis being about 3:2. *Melaleuca* of the same family has similar grains. It may be difficult to discern the characteristic polar triangle in some mounts.

Other grains with three apertures have a granular appearance. Among these are *Quercus, Fagus, Nyssa,* and *Acer negundo.* Normal grains of *Quercus* are uniform in gross morphological features, but there is a considerable range in size among the many species likely to be encountered. This range is from about 25 to 37 (but mostly around 30) microns for grains in polar view, the position in which they are usually oriented. The outer wall is rather thick and stains deeply. The granules stain darker than the general surface and seem to be randomly disposed. The perimeter of the grain in cross section is roughened by these granules. In expanded grains, the thick, colorless furrow membrane bulges considerably. The rigidity of the outer wall commonly results in a recurving at the furrow (Fig. 9-11). The furrows are of moderate length, sometimes sharply defined, but often their limits are indistinct. The polar area is of medium size, but more than that of several of the pollens with which it is commonly confused (e.g., some members of the Rosaceae). *Acer negundo* might easily be mistaken for early flowering *Quercus.* Expanded grains are usually seen in polar view and are about 27 to 30 microns in maximum dimension. The furrows are of medium length, and the polar area is probably a little smaller than in *Quercus,* but the difficulty in tracing furrows to their polar extremity makes this feature of limited diagnostic

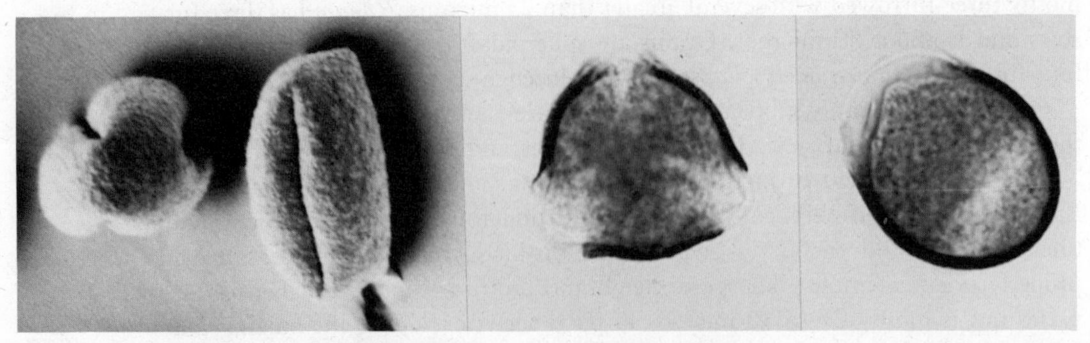

FIGURE 9-11 Quercus pollen in polar and equatorial views. The two on the left are scanning electron microscope photographs of dry grains. The two on the right are light microscope photographs of expanded grains.

value. The shape of these grains in polar view is nearly round with only a few individuals showing the recurved outline of typical *Quercus*. *Acer negundo* pollen in equatorial view is more difficult to separate from some species of *Quercus*. Granulation is variable and may be finer or coarser than *Quercus* pollen but is usually finer, and the outline is smooth without a roughened appearance in cross section. Both types of grains are somewhat flattened at the poles, appearing ellipsoidal in this view.

Some collections of *Quercus* pollen have many abortive or otherwise abnormal grains. Occasionally, field-collected samples show considerable quantities of such grains, which are often clumped and without accompanying normal pollen. These aberrant grains are commonly small, with abnormal numbers of furrows which do not expand, or no furrows at all. Without expansion, the pollen is still ellipsoidal, but with the polar axis the long dimension rather than the equatorial diameter. Granulation is often more pronounced than in normal grains, sometimes appearing as minute spines.

Fagus pollen is large (about 35 to over 40 microns); with three fairly long furrows, each with a large circular pore. The furrows do not gape widely when grains are moistened so that the grains are nearly always round. The pore is large (about seven microns) and is usually round and not sharply defined. In some collections, irregular patches of staining material cross the pore membrane like an operculum.

Nyssa pollen in equatorial view may be confused with that of *Fagus*, but the elliptical pore is distinctive with a definite thickened margin. The grains are slightly flattened at the poles, contrasting with the round grains of *Fagus*. In polar view, there are more differences; the pollen tends to be triangular, the furrow margins are thickened, and the exine in cross section is thickest midway between the apertures. There are other subtle differences of texture and aperture form.

Acer pollens other than *A. negundo* are usually described as striate or reticulate. The patterns are fairly distinctive and may be used to make at least a rough separation of species found in a given locality. In an area with a number of species flowering over a long period, *Acer* would probably be separated as striate (i.e., striate or reticulate) and nonstriate as a matter of practicality. A further separation would require high resolution microscopy and would likely be too time-consuming. It should be kept in mind that, under proper weather conditions, even insect-pollinated species such as *A. platanoides* may disperse some airborne pollen. With uniform mounting methods, *A. negundo* and *A. saccharinum* would be the common species listed as nonstriate, although faint striae may be detected in *A. saccharinum* under high magnification. *A. saccharum* may be reticulate or, as with most other species, show varying degrees of striation. Some patterns are pronounced and are frequently arranged in whorls which have been described as "fingerprint" striations. Patterning may be difficult or impossible to detect in a few grains resulting in some slight overcounting of the nonstriate category.

Expanded *Acer* pollen is usually seen in polar view. The shape is approximately spheroidal in any orientation. Grains not fully expanded may be "boat-shaped" in equatorial view. The size of normal grains may vary from about 27 to 40 microns depending on the species, degree of expansion, and orientation. Some collections, particularly of *A. rubrum*, may exhibit many abnormalities, with giants and dwarfs, and unusual numbers and forms of furrows. This is probably due to genetic complexity. Those species of *Acer* whose pollen is at least occasionally airborne do not have pores.

8. Irregularly shaped grains

These include pollens which usually are not spheroidal. Such grains commonly lack the radial symmetry of most airborne pollens, but may exhibit bilateral symmetry. Compound grains whose individual components may be spheroidal are included here.

Typha latifolia has spheroidal grains always united in tetrads. The grains may be united in any configuration; in linear, decussate, L- or T-shapes, but most commonly in square or rhomboidal groupings (Fig. 9-1). The distance between opposite sides of a square tetrad is about 31 to 41 microns. The individual components of the tetrad are a little smaller, on the average, than the single grains of *T. angustifolia*. They are similar to *T. angustifolia* in their other main features, i.e., the pitted to finely netted exine and the rather large but not sharply limited pore.

Pollen of the Juncaceae is also released in tetrads. The genera *Juncus* and *Luzula* are widespread, of common occurrence, and adapted to wind pollination, but their pollens are not abundant in air samples. Part of the reason for this is because the delicate walls easily crumple or break and, if not freshly collected, they are often unrecognizable. Unlike most monocotyledonous plants, the pollen is united in tetrahedral tetrads or with a considerable number joined in decussate form. The thin, lightly staining wall is thinnest at the distal pole of the individual components, the points most remote from the center of tetrahedral tetrads. This corresponds to the more distinct apertural area of other pollen types. For expanded grains in good condition, the juncture of the individuals is seen as a lightly stained three-armed form surrounding three globular bodies which appear granular and colorless to pale yellow. These bodies are the individual cell contents. It is frequently impossible to bring the fourth cell into the plane of focus. The juncture of the grains, the thin distal area, and other features can be more clearly seen in overexpanded mounts. A few unringed slides and some in mountants other than glycerin jelly should show pollen in various condition and may be better for reference than perfectly mounted, ringed slides. Because of distortion and breakage, pollen sizes mean little, but tetrads in good condition probably average about 40 microns in maximum dimension.

Acacia has unique compound pollen which is impossible to confuse with any other. The cells are joined in multiples of four, of various numbers for the different species, but commonly in groups of 16. The components are squarish and closely joined together. Furrows can be seen at higher magnifications. In thin mounts, the outline of the entire structure appears as a scalloped spheroid, due to the junction of the individual cells at the periphery. In thicker mounts, many grains may be turned presenting a different outline. Those which are turned ninety degrees appear much flattened and are ellipsoidal. The maximum dimension of the compound grains is about 45 to 55 microns.

Some other irregularly shaped grains show a modified ellipsoidal outline rather than the common spheroidal shape. Some of these have single furrows; others have localized wall thickenings which tend to restrict expansion to certain areas of the pollen.

Of those with single furrows, *Ginkgo* pollen is mostly broadly "boat-shaped."

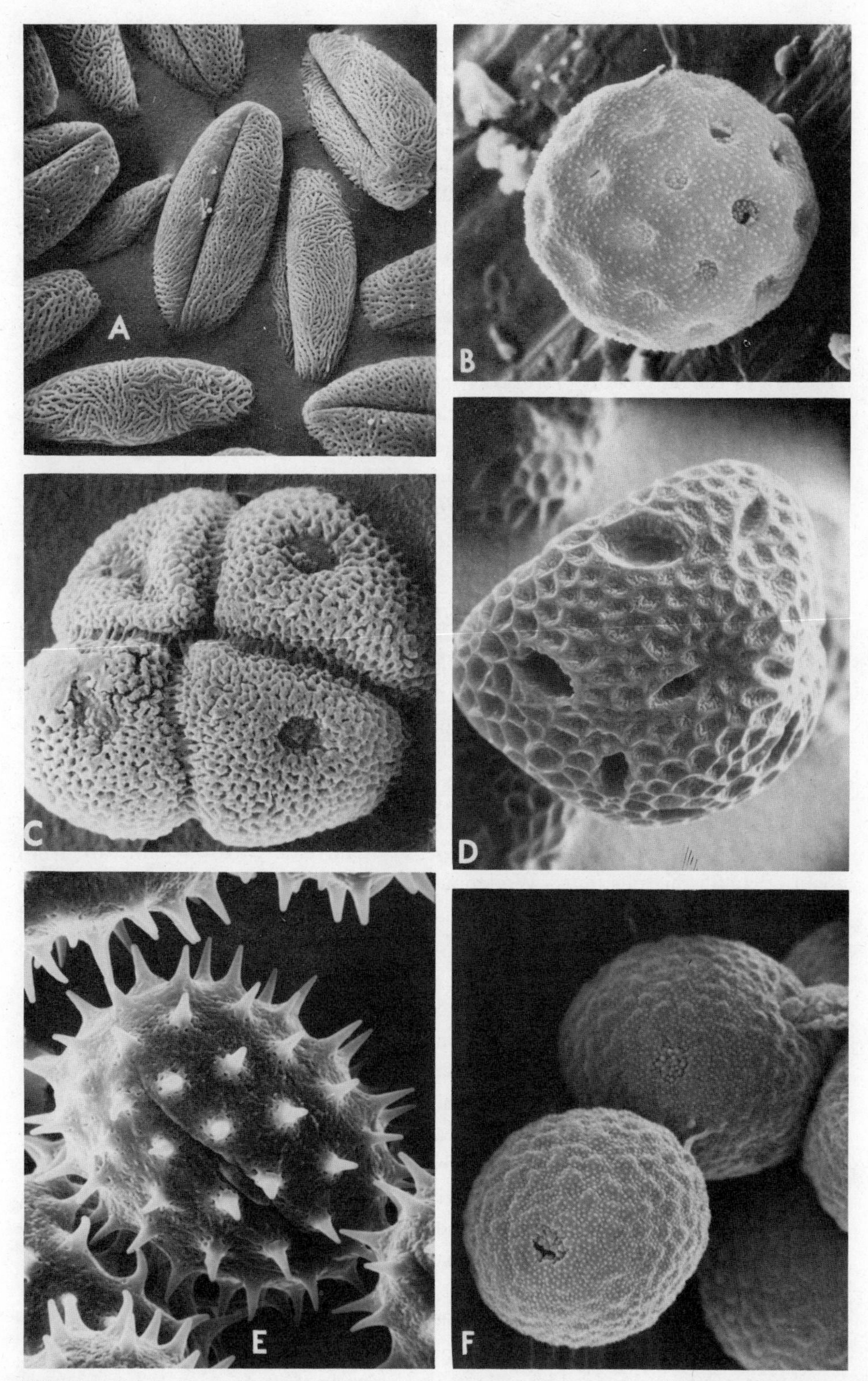

FIGURE 9-12 Scanning electron microscope photographs of pollen grains. A, striate grains of Acer grandidentata. B, multiporate grains of Allenrolfea occidentalis. C, tetrad of Typha latifolia. D, netted grains of Polygonum coccineum. E, spiny grain of Helianthus annuus. F, granular grain of Plantago major. Approximate magnifications: A, 750×; B to F, 1750×. All courtesy of Laboratory of Paleoenvironmental Studies, University of Arizona.

The exine is thin and light staining. The single broad furrow (in expanded grains) extends most of the length of the pollen and appears as an ellipse within an ellipse. The characteristic wavy exine near the furrow is usually evident if grains are oriented so that the furrow cannot be clearly seen in surface view. The exine is roughened, but is relatively featureless and would be considered smooth. The grains are about 30 microns in maximum dimension. This pollen caught on rotating impactors may cause some confusion because then they are usually spheroidal (although the exine is not ruptured or obviously distorted), and the furrow may be difficult to detect. It is likely that the furrow rim is stretched, rounding the grains and reducing contrast between the furrow area and the rest of the grain. These bear a superficial resemblance to *Populus* pollen.

Pollen of the date palm (*Phoenix*) resembles *Ginkgo* pollen. The furrow sometimes extends the entire length of the grain. Unlike *Ginkgo,* the furrow-membrane protrudes strongly in expanded grains. This appears as a broad colorless crescent in equatorial view, contrasting with the lightly stained exine. In polar view, it appears as a broad papilla amounting to about one-third of the total breadth of the grain. The furrow may be constricted at the center if grains are not fully expanded. Expanded grains are about 20 microns along their major axis while unexpanded ones are a few microns longer. The palm family (Palmae) is one of those families in which considerable diversity of pollen form is encountered (Sowunmi 1972). *Sabal* is similar to *Phoenix,* but with more apparent reticulate surface texture. The coconut (*Cocos nucifera*) has much larger pollen (over 50 microns) with a few grains having a three-split aperture, but most have a single long furrow as in *Phoenix*.

The genus *Ephedra* is of controversial gymnospermous ancestry and its pollens are unlike those of supposedly closely related plants. The grains of the various species have 5 to 15 longitudinal ridges (but no true furrows) and are from about 35 to 55 microns long (Wodehouse 1935). This "pleated" surface tends to flatten as grains expand. The pollen has a strongly elliptical outline. Expanded grains of *E. aspera* and *E. californica,* for example, have a long to short axis ratio of about 5:3.

The furrows of some pollens do not expand greatly to allow changes of volume in moistened grains, and thus they retain their ellipsoidal shape. *Castanea dentata* has small ellipsoidal grains (polar axis about 14 or 15 microns) with three long, narrow furrows. The grains are smooth. The most outstanding features are their shape; their small size; and, especially, the presence of small, elliptical, transverse furrows in the center of each of the meridional furrows. These secondary furrows are only about 2.5 microns long, and the membrane often protrudes bubblelike, giving the structure a round porelike look.

Daucus and other members of its family (Umbelliferae) resemble *Castanea* somewhat in gross characteristics. The pollen is mostly oriented in equatorial view with three long, often slitlike furrows. Each furrow has a central aperture which is variously described as a transverse furrow or elliptical germ pore (Fig. 9-5). The inner part of the exine is thicker in the vicinity of the pores, and the grains of some species are moderately constricted in this equatorial zone producing a slight dumbbell appearance. The exine is rather thick and stains deeply. The difference in staining reaction of its inner part frequently shows as a lighter equatorial band. The surface is distinctly and uniformly granular, although sometimes of minute elements. The long axis may be more than twice the equatorial diameter in some species. Some grains are large, but those taken in air samples are virtually all in the 20- to 40-micron range. The smaller grains may rarely be oriented in polar view in which case they are strongly three-angled with the furrows at the sides of the triangle.

A unique pollen type is that of the Cyperaceae. Its long triangular-, top-, or pear-shaped form is usually the basis for separating this from all other pollens. Pollen development does not parallel that of most other plants. They are considered to be modifications of the pollen mother cell with one element of the developing tetrad eventually predominating. Consequently, these pollens have been termed cryptotetrads, pseudomonads, or aberrant tetrads by different authors. The exine is thin and lightly stained. Some portions of the grain have a roughened appearance which is considered to be analogous to more clearly defined apertures of other pollens. The most distinct of these roughened areas is usually the porelike zone at the obtuse end. Lateral disjunctions of exine are considered to be of high diagnostic value for this family by Faegri and Iversen (1964). The intine is of varying thickness and is almost always much thickened (up to nine microns) at the acute end. Some grains approach the spheroidal shape and, if much crumpled, may superficially resemble *Populus* pollen (e.g., *Rhynchospora*). The size of cyperaceous pollen on field-collected samples is mostly within the 30- to 45-micron range, probably averaging about 35 microns.

References

Andersen, S.T. 1960. Silicone oil as a mounting medium for pollen grains. Dan. Geol. Unders. Raekke 4, 4(1):5–24.

Andersen, S.T. 1965. Mounting media and mounting techniques. Pages 587–598 *in* B. Kummel and D. Raup, eds. Handbook of paleontological techniques. W.H. Freeman & Co., San Francisco.

Brooks, J., P.R. Grant, M. Muir, P. van Gijzel, and G. Shaw, eds. 1971. Sporopollenin. (Proc. of a conf. held at Imperial College, London. 1970) Academic Press, New York. 718 p.

Christensen, J.E., H.T Horner, Jr., and N.R. Lersten. 1972. Pollen wall and tapetal orbicular wall development in *Sorghum bicolor* (Gramineae). Amer. J. Bot. 59: 43–58.

Cranwell, L.M. 1953. New Zealand pollen studies. The monocotyledons: A comparative account. Harvard Univ. Press, Cambridge, Mass. (Also Aukl. Inst. and Mus., Bull. 3) 91 p.

Cushing, E.J. 1961. Size increase in pollen grains mounted in thin slides. Pollen et Spores 3:265–274.

Dahl, A.O. 1969. Wall structure and composition of pollen and spores. Pages 35–48 *in* R.H. Tschudy and R.A. Scott, eds. Aspects of palynology. Wiley-Interscience, New York.

Echlin, P. 1968. Pollen. Sci. Amer. 218:80–90.

Erdtman, G. 1943. An introduction to pollen analysis. Chronica Botanica, Waltham, Mass. (Reprinted 1954 by Ronald Press, New York). 239 p.

Erdtman, G. 1952. Pollen morphology and plant taxonomy: Angiosperms. (An introduction to palynology. I.). Almqvist & Wiksell, Stockholm and Chronica Botanica, Waltham, Mass. 539 p.

Erdtman, G. 1957. Pollen and spore morphology. Plant taxonomy: Gymnospermae,

Pteridophyta, Bryophyta [Illustrations]. (An introduction to palynology. II). Almqvist & Wiksell, Stockholm and Ronald Press, New York. 151 p.

Erdtman, G. 1965. Pollen and spore morphology. Plant taxonomy: Gymnospermae, Bryophyta [Text]. (An introduction to palynology. III). Almqvist & Wiksell, Stockholm. 191 p.

Erdtman, G. 1969. Handbook of palynology. Hafner Publ. Co., New York. 486 p.

Faegri, K., and J. Iverson. 1964. Textbook of pollen analysis. 2nd rev. ed. Hafner Publ. Co., New York. 237 p.

Gullvåg, B.M. 1966. The fine structure of pollen grains and spores: A selective review from the last twenty years of research. Phytomorphology 16:211–227.

Heslop-Harrison, J. 1968. Pollen wall development. Science 161:230–237.

Heusser, C.J. 1971. Pollen and spores of Chile. Univ. Ariz. Press, Tucson, Ariz. 167 p.

Hodges, D. 1952. The pollen loads of the honeybee. Bee Research Assoc. Ltd. (London). 51 p.

Hyde, H.A. 1954. Oncus, a new term in pollen morphology. New Phytol. 54:255–256.

Hyde, H.A., and K.F. Adams. 1958. An atlas of airborne pollen grains. Macmillan, London and St. Martin's Press, New York. 112 p.

Kapp, R.O. 1969. How to know pollen and spores. Wm. C. Brown Co., Dubuque, Iowa. 249 p.

Kremp, G.O.W. 1965. Morphologic encyclopedia of palynology: An international collection of definitions and illustrations of spores and pollen. Univ. Ariz. Press, Tucson, Ariz. 186 p.

Martin, P.S. and C.M. Drew. 1969. Scanning electron photomicrographs of southwestern pollen grains. J. Ariz. Acad. Sci. 5:147–176.

Martin, P.S. and C.M. Drew. 1970. Additional scanning electron photomicrographs of southwestern pollen grains. J. Ariz. Acad. Sci. 6:140–161.

Reiter, R. 1947. The coloration of anther and corbicular pollen. Ohio J. Sci. 47: 137–152.

Samuelsson, K.E. 1965. Photomicrography of recent and fossilized pollen. Pages 626–636 *in* B. Kummel and D. Raup, eds. Handbook of paleontologcial techniques. W.H. Freeman & Co., San Francisco.

Shellhorn, S.J., H.M. Hull, and P.S. Martin. 1964. Detection of fresh and fossil pollen with fluorochromes. Nature 202:315–316.

Skvarla, J.J., and D.A. Larson. 1965. An electron microscopic study of pollen morphology in the Compositae with special reference to the Ambrosiinae. Grana Palynologica 6:210–269.

Solomon, W.R., O.C. Durham, F.L. McKay. 1967. Aeroallergens II. Pollens and the plants that produce them. Pages 340–397 *in* J.M. Sheldon, R.G. Lovell and K.P. Mathews. A manual of clinical allergy. 2nd ed. W.B. Saunders Co., Philadelphia.

Sowunmi, M.A. 1972. Pollen morphology of the Palmae and its bearing on taxonomy. Rev. Palaeobot. Palynol. 13:1–67.

Ting, W.S. 1966. Determination of *Pinus* species by pollen statistics. Univ. Calif. Publ. Geol. 58. 168 p.

Traverse, A. 1965. Preparation of modern pollen and spores for palynological reference collections. Pages 598–613 *in* B. Kummel and D. Raup, eds. Handbook of paleontological techniques. W.H. Freeman & Co., San Francisco.

Wodehouse, R.P. 1935. Pollen grains. McGraw-Hill Book Co., New York. (Reprinted 1959 by Hafner Publ. Co., New York). 574 p.

Wodehouse. R.P. 1942. Atmospheric pollen. Pages 8–31 *in* Aerobiology. AAAS Publ. 17. Amer. Ass. Advan. Sci., Washington, D.C.

Wodehouse, R.P. 1971. Hayfever plants. 2nd rev. ed. Hafner Publ. Co., New York. 280 p.

10

IDENTIFICATION OF AIRBORNE FUNGUS SPORES

Fungus spores comprise a large portion of the air flora and often outnumber pollen grains in air samples. However, they often go unidentified or unnoticed because of their small size or undistinctive shape. This Chapter illustrates a few of the most common airborne fungus spores found in samples taken primarily for airborne pollen and outlines some clues to aid in the identification of others. There will always be a number of spores which cannot be identified with certainty. Fortunately, some can be germinated and grown into mature plants on nutrient media. The mature fungus plants are much easier to identify than are the spores alone. For a practical treatment of fungi in pure culture see von Arx (1970), and for information on culture media for airborne fungi see Rogerson (1958).

Identification from spores often is possible only to the genus or family level. However, if the species can be recognized with any degree of assurance, it should be recorded because different species of the same genus may differ greatly in allergenicity or physiological properties.

The organisms grouped together as fungi are a diverse and often distantly related lot. It is difficult to make generalizations about them or even give them a precise definition. It will be useful here to divide the fungi into their major classes.

FUNGI AND OTHER ORGANISMS

If the fungi are so diverse, how can we distinguish them from other organisms? The most conspicuous feature is the lack of chlorophyll. When fungi are green, the pigment is not chlorophyll and it is not located in chloroplasts, which are absent in fungi but which are a feature of most other plants. Fungus cells are usually surrounded by a wall; this is absent in animal cells. It may be cellulose; but more often it is chitin, the substance of which insect exoskeletons are composed. Fungus cells contain mitochondria and nuclei with surrounding membranes. Bacteria, Actinomycetes, and blue-green algae do not. For further information on the Actinomycetes see Waksmann (1961), and for fungi see Alexopoulos (1962).

On a more practical level, fungus spores are usually between bacteria and pollen in size, with most of them falling in the 4–16-micron range. Most pollens, but few fungi, stain with basic fuchsin. Fungus spores also lack the complex apertures, triradiate ridges or furrows, and triporate configuration often found in pollens and spores of higher plants. Fungus spores may also become strongly pigmented as compared to pollen. Pollens, moss spores, and fern spores are mostly spherical and single-celled, whereas fungus spores may be of almost any shape or any number of cells.

SPORES AND HYPHAE

Fungus spores may be more easily understood if we describe the vegetative plant first. Most fungi have an extensive network of branched and interconnected tubes with nucleate cytoplasm inside. These tubes, called hyphae, have perforate cross walls which allow the flow of nutrients or sometimes even nuclei throughout the network. While the hyphae are growing, there is extensive biochemical activity and physical movement of the cytoplasm.

Spores usually are not produced directly on the hyphae but on some specialized structure which aids in their dispersal. Prior to the formation of the spore-bearing structure, an accumulation of nutrients occurs. The spores differ from the hyphae in being solidly packed with cytoplasm and reserve storage material, while a hypha is mostly filled with vacuoles. The spores are sealed off and there is little movement or biochemical activity. Movement of liquids in or out of these cells is also much less than in vegetative cells.

Spores have the biological functions of multiplication, dispersal, fertilization, dispersing genetic variation, and protection during unfavorable conditions. A single spore may fulfill several or all of these functions, and more than one type of spore may be produced by a single fungus. For further information on the biology of fungi see Burnett (1968).

Many spores are so adapted that they normally would not be found in air samples. Some fungi are adapted to an aquatic habitat and have flagellated spores and thin walls unable to withstand desiccation. Others are adapted to transmission from insect to insect without becoming airborne. Some are adapted to survival during dry or otherwise unfavorable conditions and remain attached to the hyphae in soil or plant remains. However, the number of fungi which produce airborne spores is sufficient to cause many problems in identification. There are more than 100,000 described species of fungi, most of which have at least one type of airborne spore. For interesting and informative reading on the dynamics of fungus spores see Ingold (1971).

SPORE MORPHOLOGY

Although fungi differ in biochemical and genetic composition, no techniques have been developed which are of practical value in using such characters on the samples that are taken primarily for pollen. This leaves only morphological characters for identification.

Since the morphology of the cytoplasm as seen with the light microscope is similar in all fungi, it is of little diagnostic value. This leaves only the spore wall for the identification of the thousands of kinds of airborne spores. This situation is not as hopeless as it may seem at first. As the aeromycologist (one who studies airborne fungi) gains familiarity with common spores, he can identify some of them as readily as he can fruits at a supermarket counter. Shape, size, and color are used as well as surface markings, number of cells, and type of appendages.

Symmetry

The study of shapes of fungus spores may logically begin with the examination of symmetry (Fig. 10-1): the comparison of shapes on opposite sides of a plane or around an axis or point. The type of symmetry gives a clue to the type of structure that bore the spore, and the classification of fungi is based, in part, on differences in spore-bearing structures. For example, spores borne in a spherical sac tend to be spherical, those in an elongated sac elongated, those squarely at the tip of a hypha exhibit radial symmetry, and those borne acentrically at the tip of a specialized cell tend to be asymmetrical.

In the categories below, spherical spores are treated under the spherical heading only and are not included under bipolar and radial, although by definition they could be included there.

Spherical spores

This type of spore is found in every class of fungi, and it is the predominant type in the groups with spores borne in a spherical structure such as the Myxomycetes and Zygomycetes (Fig. 10-2).

Bipolar symmetry

A spore which can be bisected transversely to yield resultant halves which are mirror images is said to have bipolar symmetry. This type of spore is common among endospores borne in the elongated ascus of the Ascomycetes (Figs. 10-1 and 10-6B and C). It is rare in groups with exogenous (externally borne) spores like the Basidiomycetes and Deuteromycetes, except where they are borne in chains.

Radial symmetry

A spore which can be rotated about its longitudinal axis and remain symmetrical is said to have radial symmetry. This type prevails in fungi with spores squarely on the tip of a hypha as in the Deuteromycetes and Peronosporales (Figs. 10-1, 10-3A, 10-5C, and 10-7A and C).

Helical spores

Spores which are spiral or helical in shape are restricted to the Deuteromycetes (Figs. 10-1 and 10-8E).

Asymmetrical spores

A spore which does not fit into any of the previous categories will be referred to here as asymmetrical, although it may have bilateral symmetry; which means that it will yield two mirror-image halves when bisected on one longitudinal plane. This is the type which predominates in the Basidiomycetes and is common in Ascomycetes and Deuteromycetes. It is almost nonexistent in Myxomycetes and Zygomycetes (Figs. 10-1, 10-4B and C, and 10-6D).

SYMMETRY IN FUNGUS SPORES

BIPOLAR SYMMETRY

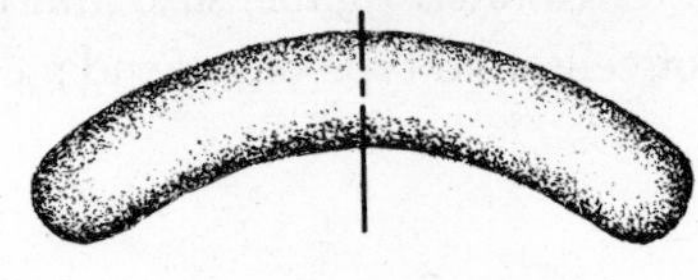

Ascospore of *Diatrype*

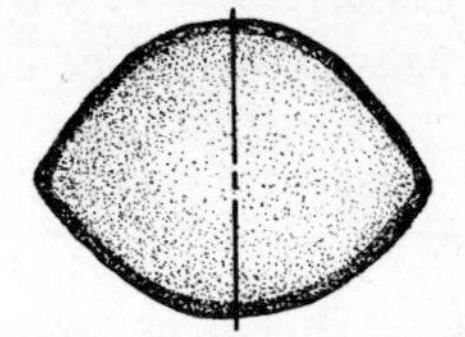

Ascospore of *Chaetomium*

RADIAL SYMMETRY

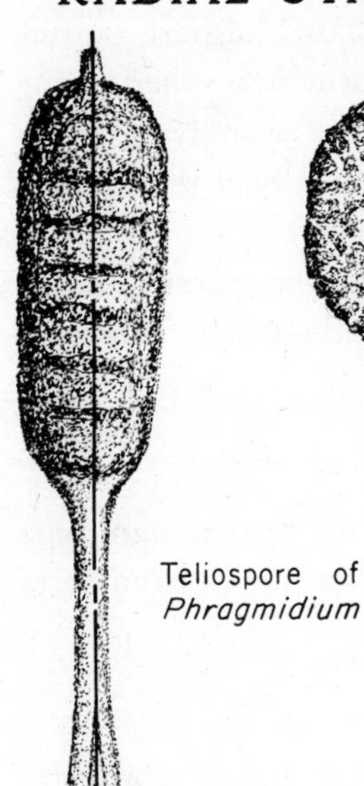

Teliospore of *Phragmidium*

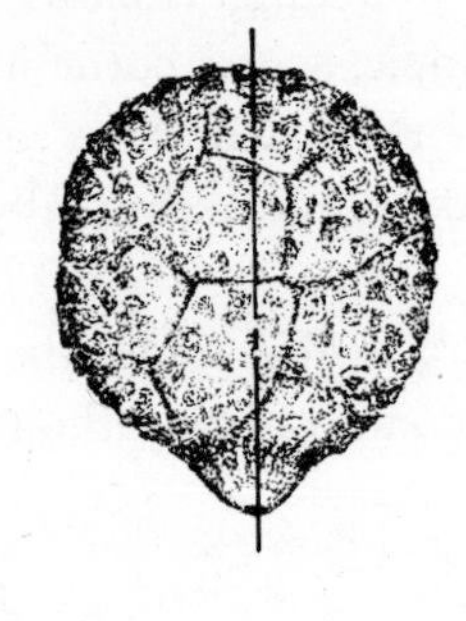

Spore of *Epicoccun* (Deuteromycete)

HELICAL SYMMETRY (Spiral)

Spore of *Helicomyces* (Deuteromycetes)

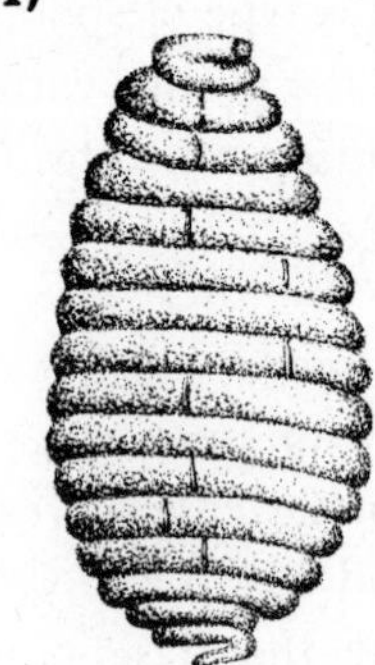

Spore of *Helicoon* (Deuteromycetes)

ASYMMETRICAL

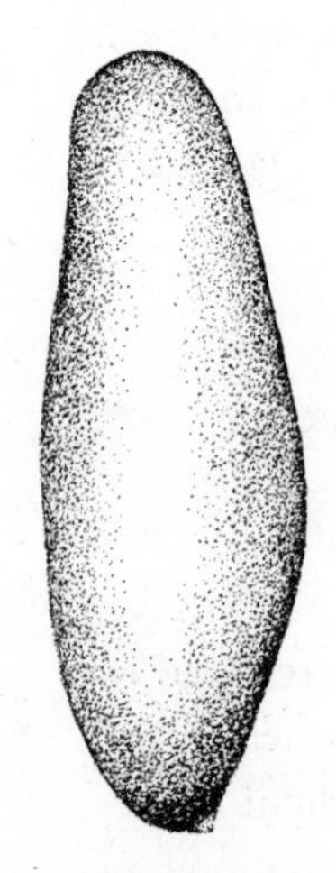

Basidiospore of *Suillus*

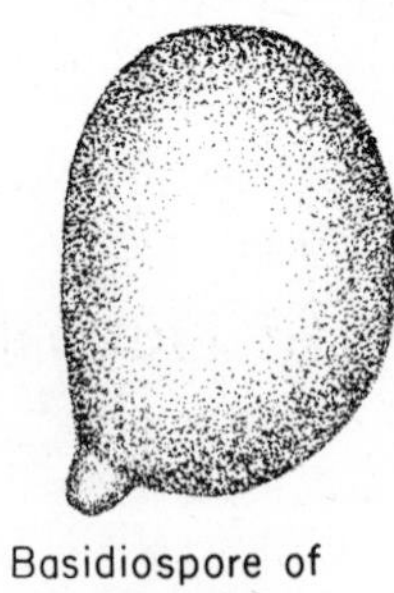

Basidiospore of *Amanita*

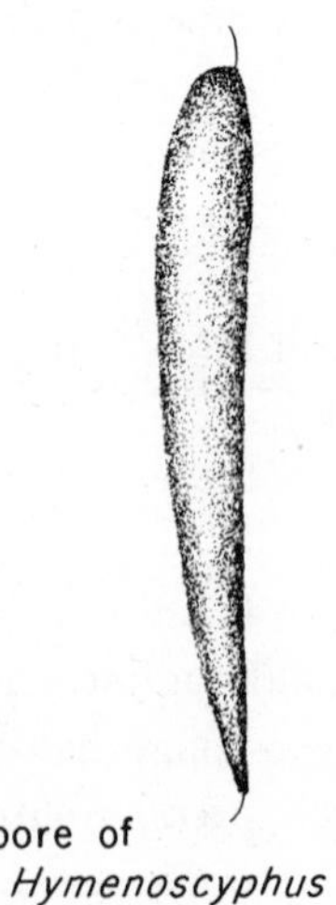

Ascospore of *Hymenoscyphus*

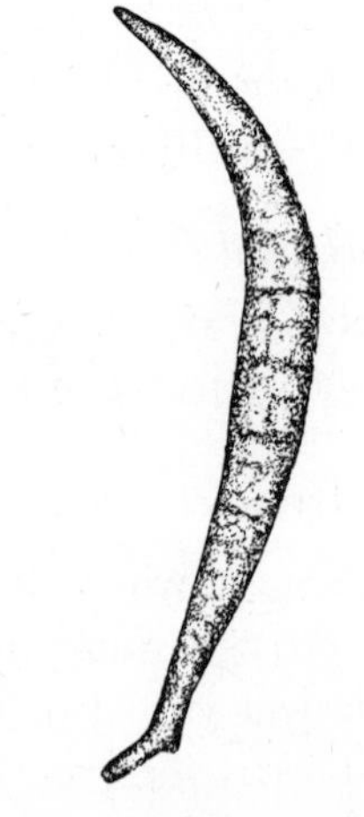

Spore of *Fusarium*

FIGURE 10-1 Symmetry in fungus spores.

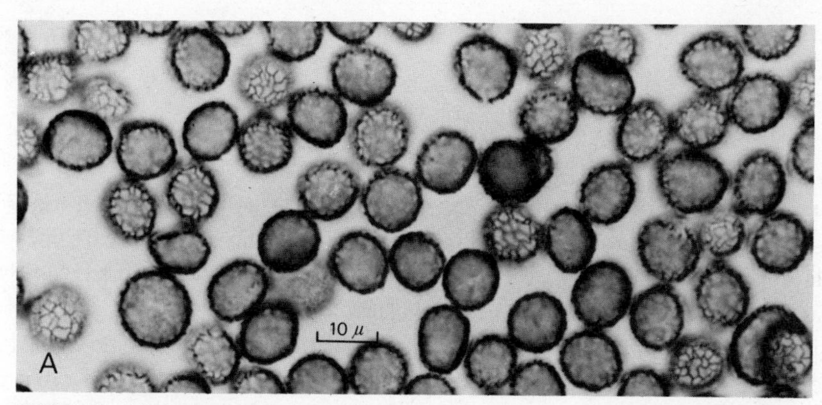

FIGURE 10-2 Spores of Stemonitis nutans.

Shape

Aside from the considerations of symmetry above, shape in fungus spores is almost as complex as the study of fungus identification itself. Specific shapes are generally useful in recognizing the genera of fungi.

Branched spores

Branching in spores occurs only in the Deuteromycetes. However, branched hyphae may be confused with branched spores. The truly branched spores are not numerous and one may become familiar with them without much trouble.

Needle-shaped spores

Spores with a length-diameter ratio of more than 10 to 1 are found in the Ascomycetes and in a few Deuteromycetes.

Angular spores

Subtle evidence of angular shape can be an important feature for distinguishing classes of fungi. Fungus spores generally are not formed as part of a tetrahedron as are spores of mosses and ferns and the pollen grains of seed plants. Therefore, the characteristic shape which forms one quarter of a sphere is not found.

Spores which are cleaved out in the interior of a sac, which are not in the form of tetrads, have an angular shape which is roughly a tetrakaidekahedron or 14-sided form. This shape is most pronounced in the teliospores of the smuts and primarily in the thicker-walled or younger spores. In the Zygomycetes, the spore cell walls are so thin that they do not hold this shape after they are released from the sporangium. Normally ascospores, basidiospores, and deuteromycete spores will not have these shapes.

Septation

Septation (the presence of cross walls) occurs in some groups of fungi (Figs. 10-6D and 10-7A, B, C, and D) but is absent in others. Every class of fungi has some species with single-celled spores, but basidiospores and spores of the Myxomycetes and Zygomycetes are rarely septate. In spores with more than one cell, usually every cell has the ability to germinate. The exception to this is some smut teliospores, where a single spherical spore or a mass of spherical spores may be surrounded by sterile cells.

Remnants of spore-bearing structures

Those spores which are borne on a specialized hyphal tip as in the Deuteromycetes, rusts, and the conidia of the Peronosporales usually have evidence (sometimes subtle) of a birth scar (Figs. 10-4B and D, 10-5B, 10-7A, B, C, and D, and 10-8B). Endospores like those of the Ascomycetes, zygomycete sporangiospores, and Myxomycetes do not have a scar or stalk. More on the type of scars is discussed under the individual groups of fungi. The teliospores of rusts are often conspicuously long-stalked and are easy to identify because of this feature (Fig. 10-5C and D).

Protrusions and appendages

An assortment of projections, lumps, or hairs is found among some fungus spores. They may arise in different manners and have different functions. Basidiospores often have a pointed outgrowth on the basal end of the spore. In ascospores similar projections may occur, but they will be at both ends of the spore and in line with the central axis (Fig. 10-1: *Chaetomium*). Also in ascospores, there may be appendages composed of gelatinous material which is best demonstrated by staining or by phase microscopy. Fine "cilia" are sometimes found on deuteromycete spores, ascospores, and very rarely on zygomycete sporangiospores (Fig. 10-1: *Hymenoscyphus*). These "cilia" are not analogous to the true cilia which are the source of locomotion in aquatic fungi. They are solid and immobile.

Pigments

A great deal has been written about pigments in fungi but, for our purposes, it is sufficient to discuss them merely in terms of the colors which they impart to the spores. Subtle differences in spore color can be used to help recognize familiar spores. It is difficult, however, to use precise color terminology for microscopic spores since different microscopes transmit light of different wave lengths. Amber, black, ochre, olive, brown, and smokey grey are colors found in ascospores, basidiospores, and deuteromycete spores. The term "hyaline" is used to describe a spore which lacks pigment.

Germ pores and slits

Many thick-walled spores have a predetermined location for a germ tube. This shows up most readily in spores with strongly pigmented walls, as under the light microscope a pore appears as a round light area (Fig. 10-4C), and a slit as a light streak parallel to the longitudinal axis of the spore or as a band around it (Fig. 10-6C). Germ pores may occur in each cell of multicellular spores, whereas the pores at the point of attachment in some Deuteromycetes are only in the basal cell (Fig. 10-7A and B).

SPORE TYPES \ CHARACTERISTICS	spherical or short ellipsoidal	bipolar symmetry	radial symmetry	helical	asymmetrical	single-celled	two or more cells	needle-shaped	branched or star-shaped	single apiculus or hilum	two apiculi	"cilia" or other appendages	long-stalked	remnant of spore-bearing structure	pigmented	hyaline	pore	germ slit	thick-walled	thin-walled	surface-ornamented	smooth	mostly over 20µ long	mostly under 20µ long	borne in chains	germinates easily in culture	distinguishing features in culture
Deuteromycete spores	⊙	Rare	⊙	⊙	⊙	⊙	⊙	Rare	⊙			⊙		⊙	⊙	⊙	⊙	⊙	⊙	⊙	⊙	⊙	⊙	⊙	⊙	⊙	sporulation
Ascospores	⊙	⊙	⊙		⊙	⊙	⊙	⊙			⊙	Rare			⊙	⊙	Rare	⊙	⊙	⊙	⊙	⊙	⊙	⊙		Some	septate hyphae
Basidiospores	⊙				⊙	⊙				⊙					⊙	⊙	⊙			⊙	⊙	⊙		⊙		Some	septate hyphae
Rust teliospores	⊙		⊙			⊙	⊙			Rare			⊙	⊙	⊙		⊙		⊙		⊙	⊙	⊙		⊙		
Smut teliospores	⊙					⊙	⊙								⊙				⊙		⊙	Rare		⊙		⊙	yeast
Rust urediniospores	⊙		⊙			⊙							⊙	⊙	⊙	⊙			⊙		⊙			⊙			
Zygomycete sporangiospores	⊙					⊙									⊙	⊙				⊙	⊙	⊙		⊙		⊙	robust hyphae
Peronosporaceae and Pythiaceae (conidia)	⊙		⊙			⊙				⊙				⊙		⊙			⊙			⊙	⊙			Some	aseptate hyphae or zoospores
Myxomycete spores	⊙					⊙														⊙	⊙			⊙			
Albuginaceae (conidia)	⊙	⊙	⊙			⊙								⊙		⊙			⊙			⊙		⊙	⊙		
Gasteromycete basidiospores	⊙					⊙				⊙				⊙	⊙	⊙			⊙	⊙	⊙	⊙		⊙			septate hyphae

Note: ⊙ *indicates that the characteristic may occur. A blank indicates that the characteristic is NOT present.*

Summary of Spore Characteristics in the Fungi

Surface ornamentation

The surface of fungus spores may be ornamented with ridges, furrows, bumps, spines, or pegs, often with much variation among genera or even species. Electron microscopy, particularly the surface replica technique with the transmission electron microscope, demonstrates that many spores thought to be smooth are minutely ornamented. Similar ornamentation may be found in spores of very diverse groups of fungi. This makes it risky to place too much emphasis on this character in identifying spores.

Surface ornamentation usually is formed by the partial breakdown of outer wall layers or by differentiated outgrowths of inner wall layers. In either case, different wall layers are exposed on the spore surface. For this reason, it is sometimes possible to use stains to accentuate ornamentation for light microscopy. For techniques see Pegler and Young (1971). Several aspects of spores change with variations in environment during their maturation process. Ornamentation is one character likely to be affected.

CLASSES OF FUNGI

The names of fungus classes are given by the scientific name and also by the English common name where it is applicable. In some cases, there is no suitable common name. Scientific and common names may be used interchangeably. For a quick reference to the names of fungus genera see Ainsworth (1971).

Myxomycetes (slime molds)

This class, composed of about 400 species, has unprotected protoplasm in place of hyphae in the vegetative stage. The spores are borne in stalked heads covered by a membrane. These spores, which are all wind-dispersed, are cleaved from the cytoplasm and are surrounded by a wall of chitin or cellulose. The spores are always one-celled, spherical or nearly so, and are usually ornamented, pigmented and are 6–18 microns in diameter. They have no stalks, pores, points of attachment, nor appendages (Fig. 10-2). Sometimes the young spores retain vestiges of the tetrahedral shape in which they are often formed or they may have more pronounced ornamentation on one quarter of their surface. In culture, spores seldom germinate; when they do, they produce amoeboid cells or germ tubes.

It is difficult to distinguish some myxomycete spores from some smut spores. Familiarity with the common spores from each group is the best way to tell them apart. For further information on Myxomycetes see the treatment by Martin and Alexopoulos (1969).

Oomycetes

This primitive group of fungi is characterized by relatively wide hyphae, 7–20 microns in diameter, with very few cross walls, motile zoospores with two flagella, cellulose cell walls, and sexual reproduction involving a saclike female gamete and male gametes which are differentiated hyphal branches. This class has both aquatic and terrestrial forms. The latter are in the order Peronosporales, with about 250 species, including *Albugo* (white rusts), *Peronospora* (downy mildews), and

Phytophthora of which the potato late blight fungus is an example. Members of the Peronosporales are plant parasites or soil saprophytes. They have sporangia, most of which are capable of producing zoospores, but which also can germinate with a germ tube like the spores of other fungi (Fig. 10-3A). These sporangia, which are referred to as conidia, are easily detached or sometimes forcibly released from the hyphae in large numbers. The spores may become very abundant during periods of high humidity and epidemics may occur in the case of the parasites. The conidia are larger than most fungus spores and may reach 40×60 microns. They are hyaline, usually thick-walled, smooth, single-celled, and radially symmetric. The families Peronosporaceae and Pythiaceae have solitary conidia, often with an apiculus at the apex and a narrow remnant of the spore-bearing hypha at the base. The family Albuginaceae differs in bearing conidia in chains. The conidia are barrel-shaped with both bipolar and radial symmetry. Before germination, however, they may become pear-shaped.

Zygomycetes

The class Zygomycetes (about 250 species) has robust, latently septate hyphae, and a sexual stage characterized by an opaque, thick-walled structure (the zygospore) formed by the fusion of two specialized mycelial branches. No motile spores are formed but abundant airborne asexual spores are. These spores are borne in a spherical saclike sporangium which is supported on a stalk. With the exception of some species where the sporangium contains only one spore, the spores are similar. They are thin-walled, hyaline or slightly pigmented; without any differentiated point of attachment, pores, or germ slits; and are spherical or ellipsoidal, always one-celled, 2–18 microns (mostly 4–8 microns), smooth, or minutely spiny, or with longitudinal ridges.

In nature, most of these organisms are soil saprophytes which come to the attention of man by causing food spoilage and laboratory contamination. In culture they grow very rapidly, fruit readily, and are not too difficult to identify.

There are two genera commonly found in air samples: *Rhizopus* (Fig. 10-3B) which releases its spores dry and is most abundant in dry weather and *Mucor* which forms its spores in drops of slime and is adapted to insect and wet weather

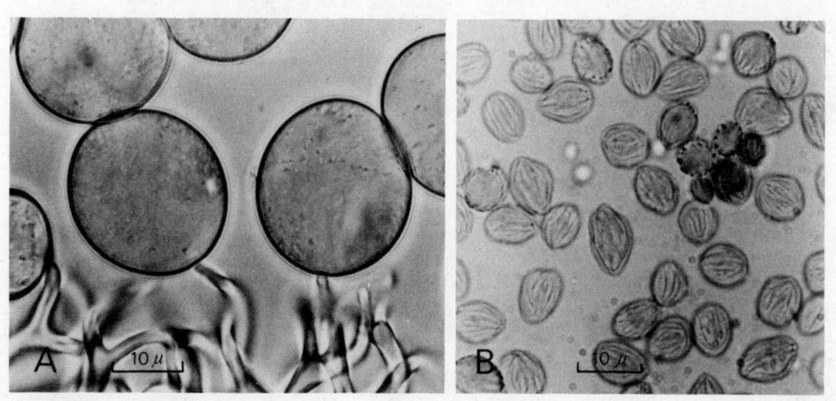

FIGURE 10-3 A, Plasmopara viticola conidia attached to the hyphae. B, Rhizopus arrhizus.

dispersal. See Zycha et al. (1969) and Kramer et al. (1960*a*) for further information.

Basidiomycetes

There are about 12,000 species in the class Basidiomycetes which is distinguished by the basidium, a club-shaped cell on which the sexual spores are borne externally (Fig. 10-4A). They also have frequently septate hyphae which seldom exceed 8 microns in diameter and which sometimes have a unique bucklelike structure (clamp connection) at the septum. Since the class is large and diverse, some common orders are here treated separately.

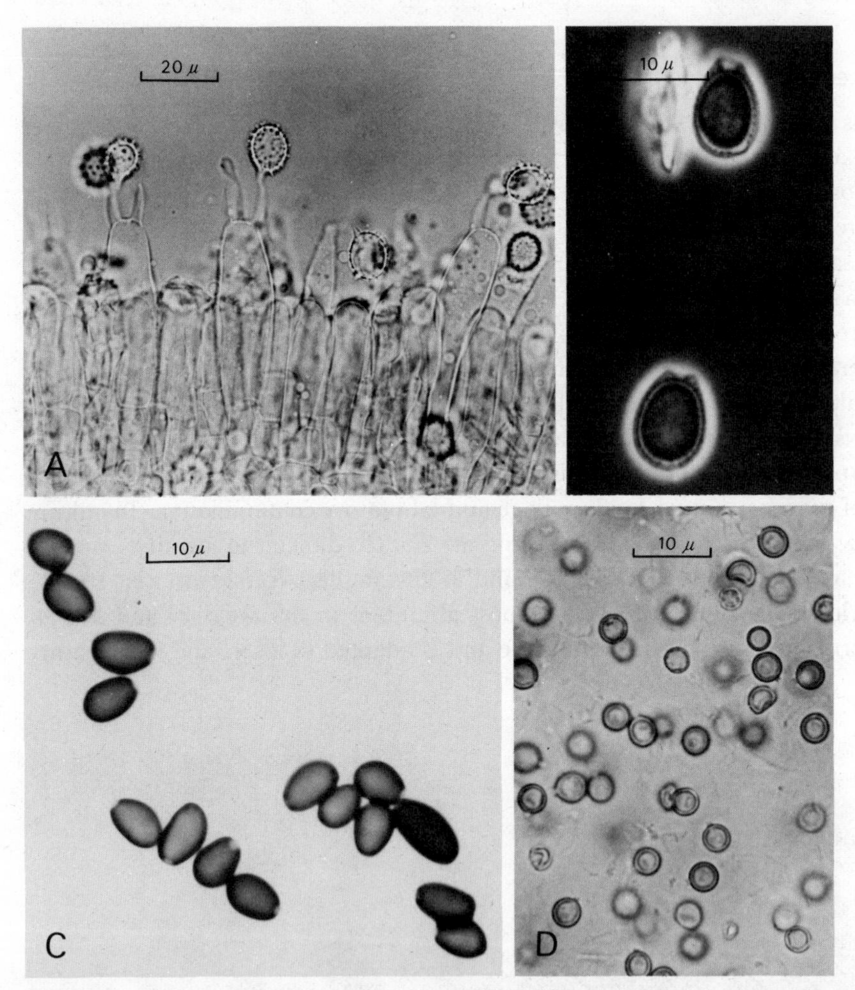

FIGURE 10-4 Spores of mushrooms (Basidiomycetes). A, Russula delica basidiospores and basidia. B, Ganoderma applanatum basidiospores with a pore surrounded by a collar at the upper end and a thin spot in the wall at the other end where it was attached to the basidium. C, Coprinus micaceus basidiospores with a light-colored germ pore at one end. D, Calvatia gigantea basidiospores with a small remnant of sterigma attached, which is typical of the Gastromycetes.

Ustilaginales (smuts)

The Ustilaginales (about 700 species) are plant parasites that produce basidia on a teliospore or "smut spore." These teliospores are produced in closepacked masses on the host tissue. They are spherical, mostly 5–14 microns in diameter, yellow to brown under the microscope but appear brown to black in dense masses. They do not have birth scars but are often slightly angular and may adhere to one another. They have no germ slits or pores. Usually, they are ornamented and thick-walled (Fig. 10-5A).

The basidiospores (produced on the basidia which are in turn produced from the teliospores) are asymmetrical, elongate, slender, and hyaline. In culture they often form yeastlike colonies. See Fischer (1953) for identification of smuts and Pady and Kramer (1960*a*) for further information on airborne smut spores.

Uredinales (rusts)

The Uredinales (about 4,600 species), which include some of our most destructive plant parasites, produce as many as five different kinds of spores in a single species. Only two are important as airborne spores: the urediniospores which are dispersed in great quantity in the spring and summer and the teliospores which are usually produced in the summer or fall and often function as resting spores. The urediniospores are single-celled, spherical or elliptical, spiny, moderately thick-walled, light straw-colored under the microscope, and attached by a long stalk which is usually broken off from those spores that become airborne (Fig. 10-5C). Urediniospores are very similar in different species and even in different genera. The genera of rusts are based on teliospore characters which vary tremendously throughout the group. Teliospores are generally thick-walled, radially symmetric, long-stalked, brown-colored, and with one or more germ pores per cell. They are often multicellular, ornamented, and may have an apiculus or appendage at the apex. They are among the largest of fungus spores and range from 25 to over 100 microns in length (Fig. 10-5B and D). Rust spores generally do not germinate nor grow readily in culture. See Cummins (1959) for identification of rust genera.

Agaricales (mushrooms and their relatives)

This is the largest group of the Basidiomycetes (about 7,000 species) and the most difficult to identify in airborne samples. Spores produced by this group are basidiospores borne on fruiting bodies, of which the mushroom is a familiar example. The spores are seasonal, being most prevalent in the fall; and periodic, being most numerous in the early morning hours. Basidiospores are recognized by their asymmetrical shape and acentric protrusion (hilum) at one end. This hilum, which is sometimes incorrectly referred to as an apiculus is composed of the same material as the cell wall (Fig. 10-1: *Amanita* and *Suillus*). It usually projects several microns, but in some species it is reduced to a thin spot in the cell wall (Fig. 10-1: *Ganoderma*). Most basidiospores are single-celled, range from spherical to slightly elongate, and are 5–15 microns in length. Germ pores are common opposite the hilum on dark-colored spores, but germ slits are not known. Spore color has been used as a main feature in classification schemes for mushrooms since very early times, but often it is necessary to have heavy deposits of spores to

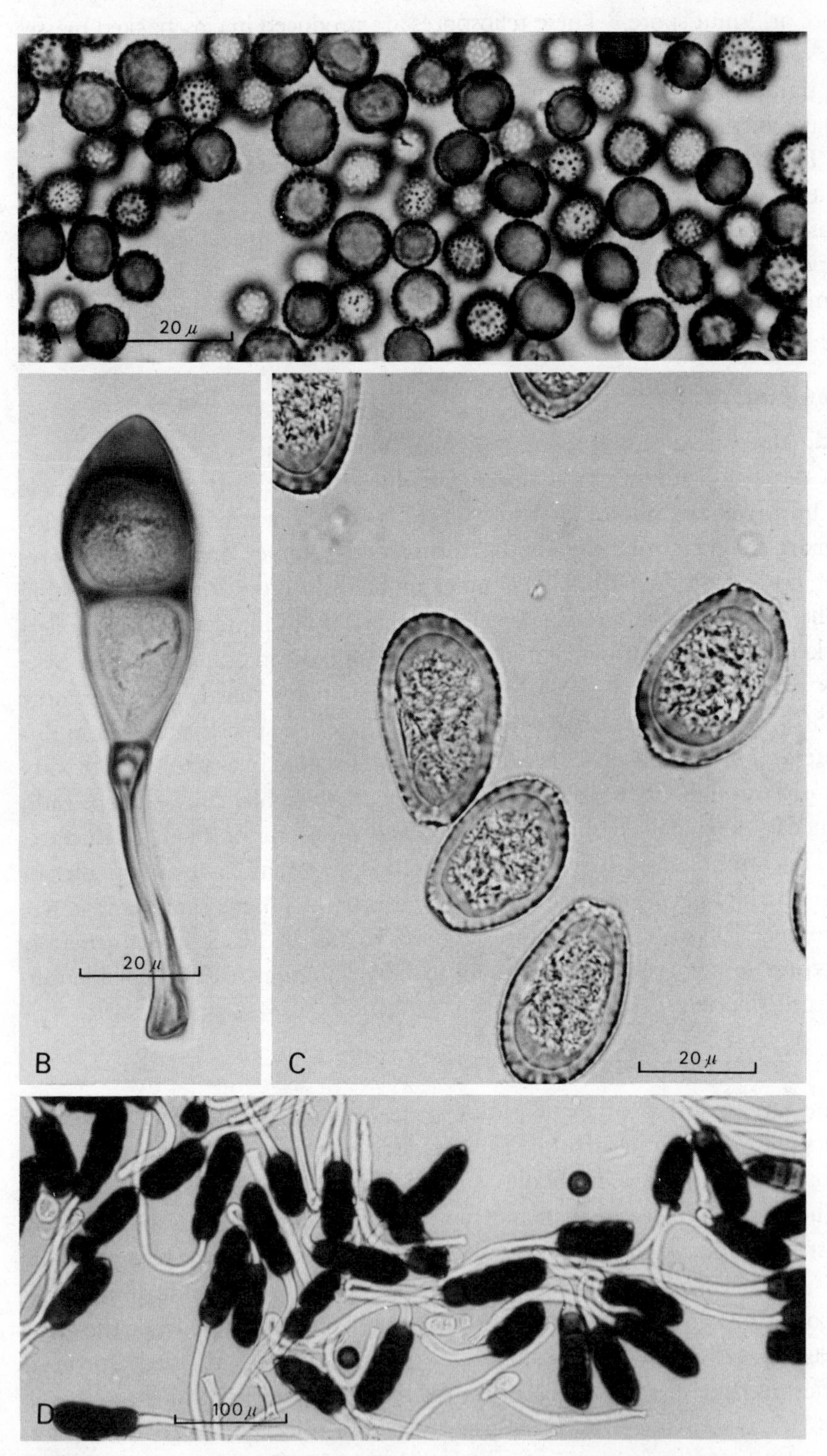

FIGURE 10-5 Spores of smuts and rusts (Basidiomycetes). A, Ustilago maydis teliospores. B, Puccinia graminis teliospore. C, Puccinia graminis urediniospores. D, Phragmidium potentillae teliospores.

distinguish subtle shades in spores which appear colorless under the microscope. Many mushrooms are difficult to identify even when the entire plant is available; identification from spores alone is often impossible. The aeromycologist must resort to categories such as "hyaline basidiospores" or "brown basidiospores." Several families of Agaricales are recognizable from their spores. These are: Rhodophyllaceae with angular spores, Boletaceae with spores two to three times as long as broad (Fig. 10-1: *Suillus*), Russulaceae with raised reticulate ornamentation that is stained blue in iodine, and Coprinaceae having nearly black spores with a conspicuous pore (Fig. 10-1: *Coprinus*), to name a few. For references see Pegler and Young (1971) and Horak (1968).

Gasteromycetes (puffballs and earthstars)

This series of Basidiomycetes (about 700 species) differs from others in not releasing its basidiospores directly into wind currents but by holding the spores in a closed fruit body. These spores are released later by puffing, although they are not forceably shot from the basidium. They do not have a hilum but they do retain a short segment of the sterigma which connects the basidium and spore. They are spherical instead of asymmetrical like most other basidiospores (Fig. 10-4D). Gasteromycetes tend to release spores later in the season than the mushrooms. Their spores may be very abundant locally since a single fruit body of a puffball may produce millions of spores.

Ascomycetes

The distinguishing feature of the Ascomycetes, which include about 15,000 species, is the sexual spores borne in a specialized sac: the ascus (Fig. 10-6A). This sac is always composed of a single cell and often has a mechanism to release the spores, which are usually eight in number. The types of spore release mechanisms and the types of fruiting bodies are the basis of classification. Most Ascomycetes, with the exception of the truffles which grow underground, have airborne spores. Many have airborne spores in addition to their ascospores. These other spores are not the result of meiosis nor sexual fusion and are known as imperfect spores. This stage of the life cycle is treated separately under the class Deuteromycetes.

Ascospores, naturally, demonstrate no evidence of having been attached to a hypha since they are endogenously formed. They are very diverse in their shapes but are never helical nor branched and seldom perfectly spherical. They may be very long and narrow. Most bipolar spores such as the allantoid or sausage-shaped spores are ascospores. Even more common is a straight or slightly curved spore which is tapered more at one end than at the other. This tapered end is the lower end of the spore as it lies in the ascus and trails behind the spore as it is squeezed through the opening at the tip of the ascus and projected through the air. Septation and pigments are also diverse and "cilia" and apiculi sometimes occur. Ascospores may be distinguished from basidiospores because the latter have an acentric hilum at one end only, whereas ascospores either have one squarely at each end or none at all. Since the ascomycetes are such a large and varied group, only a few are discussed here. For a more complete treatment see Dennis (1968).

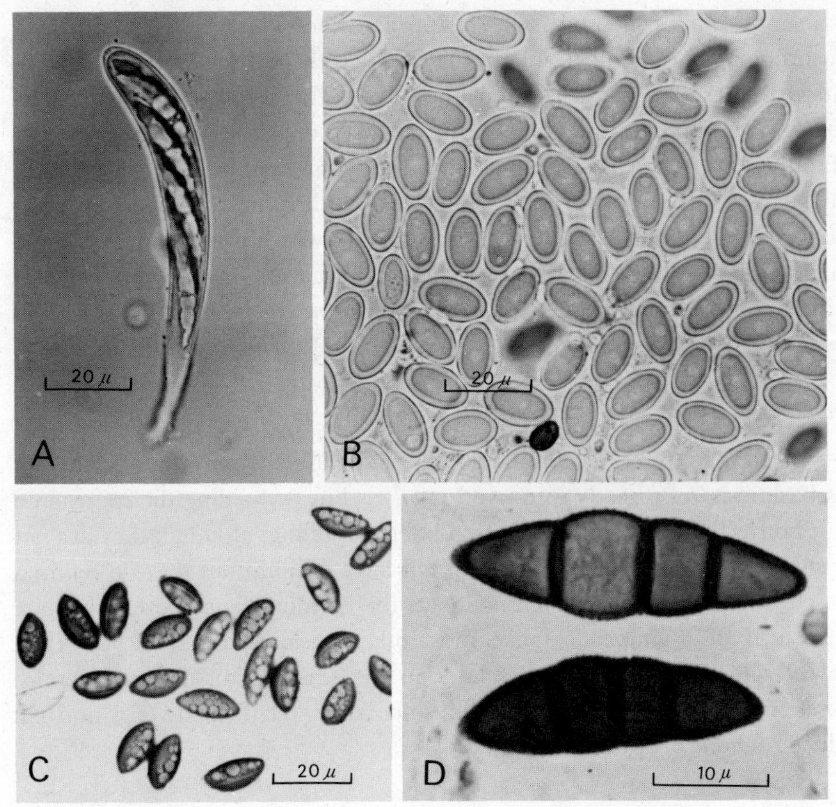

FIGURE 10-6 Spores of Ascomycetes. A, Hymenoscyphus caudatus ascus with ascospores enclosed. B, Peziza repanda ascospores. C, Daldinia concentrica ascospores with a minute longitudinal germ slit. D, Leptosphaeria doliolum ascospores.

Pezizales (the large cup fungi)

This group of about 500 species is distinguished by the type of spore release mechanism at the tip of the ascus. It is in the form of a small lid or flap. The fruit bodies are shaped like a cup or a stalked head. The spores are large and mostly between 15 and 25 microns in length. In the common airborne species, they are usually hyaline and are often ornamented with bumps or reticulate ridges. They have bipolar symmetry, are spherical, oval or ellipsoidal, seldom more than twice as long as broad, and always one-celled (Fig. 10-6B). These fungi are seasonal and in most areas occur in the spring.

Helotiales (small cup fungi)

This group of about 1,500 species is characterized by having a small plug at the tip of the ascus and by having fruit bodies which resemble tiny goblets or saucers. Their spores are generally 10–15 microns long and usually more than twice as long as broad. They are hyaline, one-celled, smooth, and slightly curved along the longitudinal axis. Most are also more slender at one end. The Helotiales are not as seasonal as the Pezizales and may be found at any time unless the weather is below freezing or very dry (Figs. 10-1: *Hymenoscyphus* and 10-6A).

Leptosphaeria

This genus (about 200 species) is very common in air samples. Its spores are usually yellow to brown, of four or more cells, slightly curved longitudinally, and often with one cell on the upper part slightly larger than the others. These fungi are very common on old grass culms and other plant stems. Their spores may be encountered at any time of the year (Fig. 10-6D).

Xylariaceae

The spores of this ascomycete family are almond-shaped, single-celled, dark brown to black, and with a germ slit (Fig. 10-6C: *Daldinia*). They are primarily wood saprophytes, found in wooded areas throughout the world but most abundant in the tropics.

Diatrypaceae

The spores of this family are small, single-celled, hyaline to light brown, and sausage-shaped. These fungi are saprophytes of wood or bark and are common near wooded areas in the spring and fall (Fig. 10-1: *Diatrype*).

Deuteromycetes (Imperfect Fungi or Hyphomycetes)

This group of about 15,000 species is composed of the asexual stages of Ascomycetes plus many fungi in which no sexual stage is known. They have the greatest diversity of spore shape. They produce most of the common airborne fungus spores.

All helical and branched spores belong to this class, but long needle-shaped spores and spores with bipolar symmetry are rarely found here. Radially symmetric spores with a small birth scar at the base are the most common. This scar differs slightly from that of the basidiospores in having a small, round, flat area where the hypha was attached or by retaining a remnant of the hypha (Fig. 10-8B, C, D, and G). The point of attachment in deuteromycete spores is much broader than the hilum of basidiospores, which appears as a point under the light microscope (Fig. 10-4A and B). Fortunately, many deuteromycetes grow and produce spores in culture. This makes identification much easier.

Many classification schemes have been devised for this group. The earliest of these separated them by the type of fruiting body produced, spore color, shape, and septation. This artificial system sometimes places unrelated species together, but for the aeromycologist who works at identifying fungi from spores only, it is very useful. Modern classifications are based on the method by which the spores are formed on the hyphae. This scheme is better at grouping related fungi, but is difficult to use if one does not have the whole plant. For a good review of these classifications see Barron (1968). For identification of deuteromycetes see Barnett and Hunter (1972) and Ellis (1971).

Porosporae

In this group, spores are formed through pores in the conidiophores (spore-bearing hyphae). These spores are usually pigmented, large, thick-walled, and often septate. The base always has a simple pore which is closed off. This pore should not be mistaken for a germ pore; it occurs at the base of multicellular spores instead of in each cell. Some very important airborne spores in this group include: *Alternaria* (Fig. 10-7D), *Curvularia* (Fig. 10-7B), *Drechslera* (Fig. 10-7A), and *Torula*.

Alternaria alternata (usually referred to as *Alternaria tenuis* in the older literature) is one of the two most conspicuous and prevalent fungi in the air (Fig. 10-7D). It is a cosmopolitan saprophyte of plant remains and almost any other available substrate such as: food, cloth, or damp walls of houses. It is a difficult matter

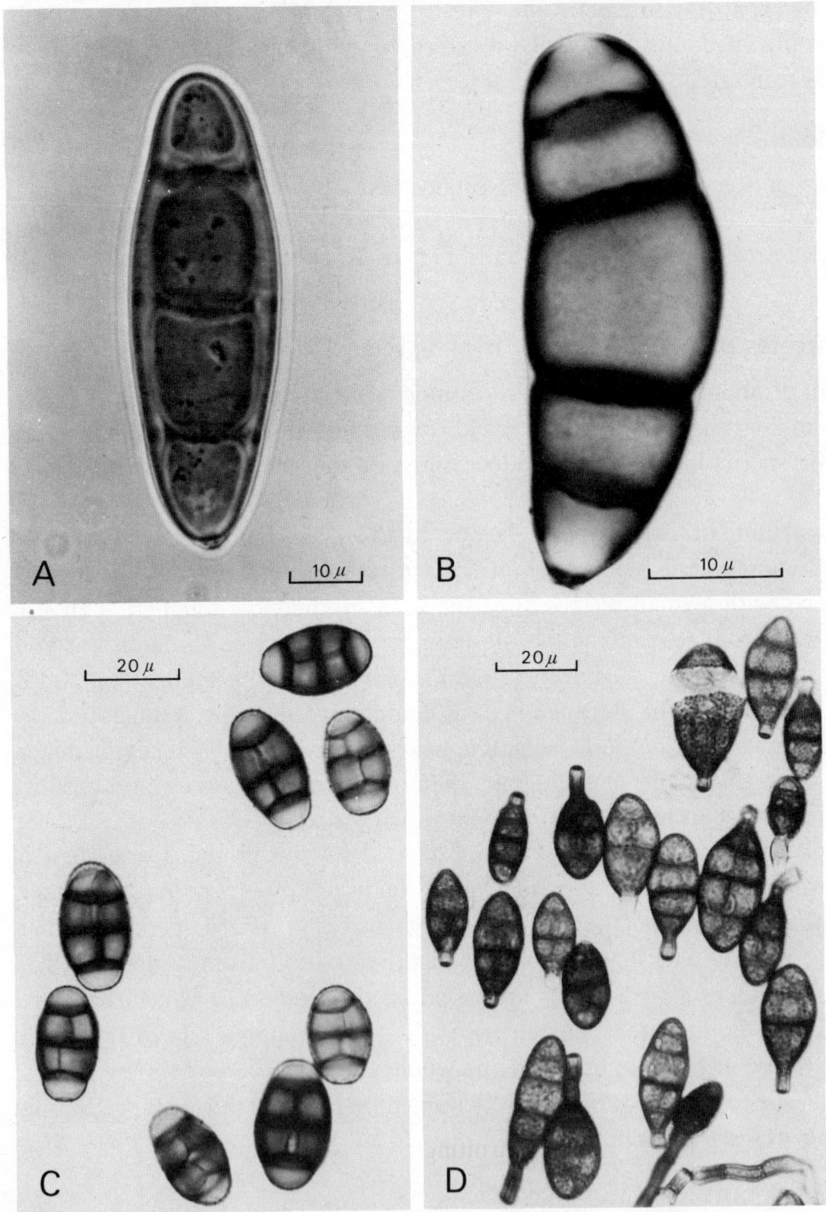

FIGURE 10-7 Spores of Deuteromycetes. A, Drechslera ravenelii spore. B, Curvularia fallax spore with pore at lower end which is typical of the Porosporae. C, Pithomyces chartarum spores. D, Alternaria spores.

to separate the approximately 40 species of *Alternaria* from the spores alone, since many species are restricted to certain host plants and the spores are only slightly different. The spores of all species of *Alternaria* are subject to great variation with relative humidity, light, temperature, and nutrition. For further information see Kramer et al. (1959*b*).

Phialosporae

This group has spores formed on specialized cells called phialides. They are usually vase-shaped and often flared at the tip. The spores are formed and released from the tip in succession and often form long chains. These spores are often difficult to identify as they tend to be small, hyaline or light-colored, single-celled, spherical or ellipsoidal, and often without a trace of a birth scar. They are easily mistaken for small ascospores.

The two most important genera in this group are *Aspergillus* (Fig. 10-8A) and *Penicillium*. Both genera contain a large number of species which are difficult to distinguish in culture or on natural substrate and are nearly impossible to distinguish from spores alone. Some success has been attained with electron microscopy (Hess et al. 1968), but the techniques are time-consuming and expensive and therefore are not yet of practical value. The spores of both genera tend to be faintly pigmented under the microscope but some shade of blue, dark green, black, or brown in mass. They are always single-celled, nearly spherical, and are often found in short chains. These fungi are saprophytes on almost any kind of organic debris and demonstrate great physiological diversity. They produce asexual spores in large dry masses which are passively dispersed in dry conditions. They grow and develop quickly but also dry up quickly as the substrate dies. These airborne spores may be encountered throughout the year in both indoor and outdoor air, but they are most common in outdoor air in the spring and fall. Airborne spores from *Aspergillus fumigatus,* one of the commonest species, are known to cause infection in humans and animals, especially in poultry. For further information on these genera see Raper and Fennell (1965) and Raper and Thom (1949).

Another important genus usually included here is *Fusarium* (Fig. 10-8B) whose species are soil saprophytes or plant pathogens. They grow readily in culture and are recognizable by their cottony white colonies which diffuse a pink-red pigment. The asexual spores are borne in slimy masses and are readily dispersed during wet periods. It is suggested by Kramer and Pady (1960*b*) that the spores rapidly lose viability in the air. For taxonomic references, see Booth (1971) and Toussoun and Nelson (1968).

Other Deuteromycetes

Most of the Deuteromycetes other than the Porosporae and Phialosporae are difficult to place in a series on the basis of the spores alone, but many have distinctive spores by which they can be identified to genus or species.

Cladosporium (some of which were formerly called *Hormodendrum*) is the most common airborne fungus spore in most areas of the world. It can be recognized easily from the spores which are ellipsoid to cylindric, lightly melanized, and with a darker protuberance at one or both ends. The spores are borne in branched chains, and the central cells of the chain are longer than the others. The

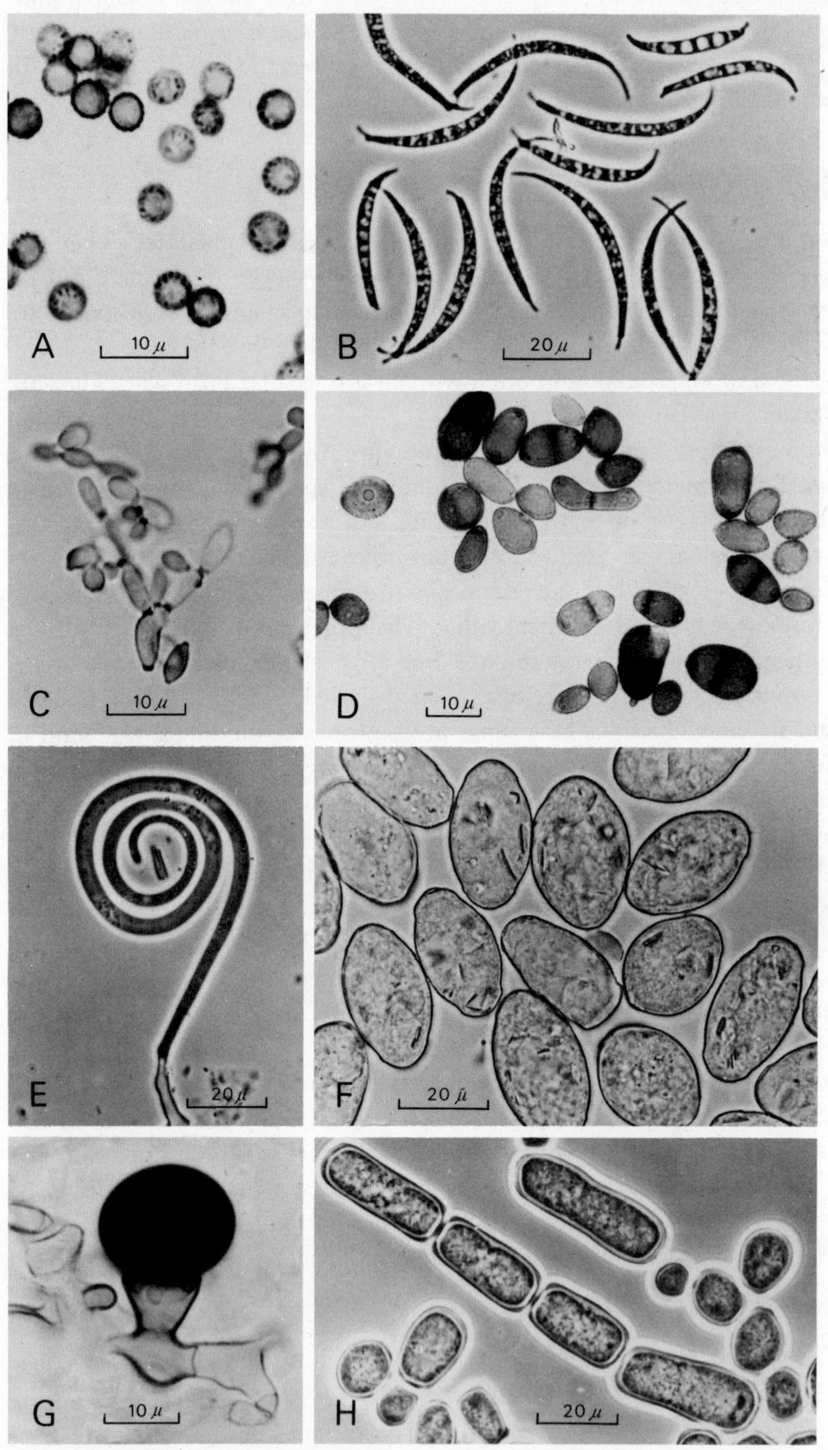

FIGURE 10-8 Spores of Deuteromycetes. A, Aspergillus niger. B, Fusarium. C, Cladosporium cladosporioides. D, Cladosporium herbarum. E, Helicomyces roseus. A single coiled spore still attached to the hypha. F, Erysiphe graminis conidia. The irregularity in outline near the ends is typical for these spores. G, Nigrospora sphaerica. A single spore attached to the hyphae. H, Monilia.

spores at the branches of the chain have three dark scars instead of two and those at the tip have only one. Often the entire chain becomes airborne and is found intact on sampler slides. The unit acts as one multicellular spore for starting a new colony.

Cladosporium is saprophytic on plant debris and is sometimes a secondary parasite of plants. The spores are produced dry but are found in greatest quantity after a period of rainfall. The season of highest occurrence coincides with the crop growing season, but spores can be found throughout the year. Diurnal fluctuation is noticeable with the highest number of spores in the air in the early afternoon. Several of the common species can be identified by size, shape, and septation on sampler slides. The most common are *Cladosporium cladosporioides* (Fig. 10-8C) and *Cladosporium herbarum* (Fig. 10-8D). For further information on *Cladosporium* see de Vries (1952) and Kramer et al. (1959*a*).

Sporobolomyces (Fig. 10-9A) is a yeast which at different times has been in cluded in different classes of fungi. It grows as colonies of single cells, but some species have the ability to form short hyphae. It differs from other yeasts by having a spore discharge mechanism like the basidiomycetes but apparently with no sexual stage. The colonies are recognized in culture by being pink and having a frosty appearance due to the spores. The spores are very tiny and resemble small basidiospores. The vegetative cells may also become airborne and resemble most other yeast cells in being ellipsoidal, but occasionally the distinctive spore-bearing cell is found which has an elongated sterigma on one side. *Sporobolomyces* commonly grows saprophytically on leaf and flower surfaces and is often overlooked on sampler slides because of its small size.

Aureobasidium pullulans (Fig. 10-9B) is another deuteromycete with a yeastlike habit. In culture it forms a distinctive slimy gray colony with a black center. The mycelium of this colony is hyaline at first but becomes black with age. The spores are formed from small bumps along the hyphae. It may occur in nature on almost any substrate which is habitable by fungi, and it is a frequent inhabitant of damp houses. It is usually difficult to identify on sampler slides but is unmistakable in culture. For a comprehensive taxonomic study of the yeasts see Lodder (1970).

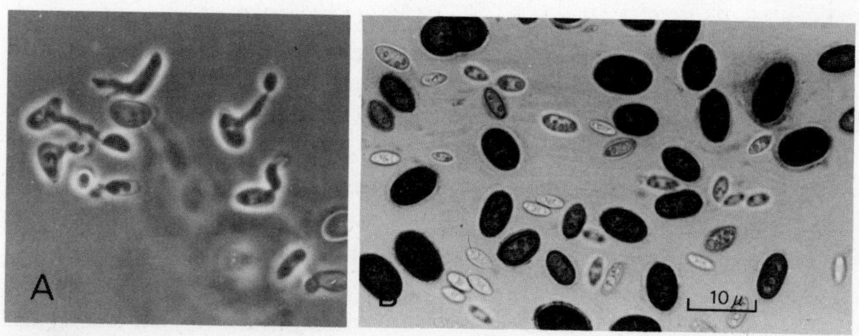

FIGURE 10-9 A, Sporobolomyces. Yeastlike cells bearing spores. B, Aureobasidium pullulans. This species has different kinds of cells which may become airborne.

References

Ainsworth, G.C. 1971. Ainsworth and Bisby's dictionary of the fungi. 6th ed. Commonw. Mycol. Inst., Kew, Surrey. 663 p.

Alexopoulos, C.J. 1962. Introductory mycology. 2nd ed. John Wiley & Sons, New York. 613 p.

Arx, J.A. von. 1970. The genera of fungi sporulating in pure culture. J. Cramer, Lehre, W. Germany. 288 p.

Barnett, H.L., and B.B. Hunter. 1972. Illustrated genera of imperfect fungi. 3rd ed. Burgess Publ. Co., Minneapolis, Minn. 241 p.

Barron, G.L. 1968. The genera of Hyphomycetes from soil. Williams & Wilkins Co., Baltimore, Md. 364 p.

Booth, C. 1971. The genus Fusarium. Commonw. Mycol. Inst., Kew, Surrey. 237 p.

Burnett, J.H. 1968. Fundamentals of mycology. St. Martin's Press, New York 546 p.

Cummins, G.B. 1959. Illustrated genera of rust fungi. Burgess Publ. Co., Minneapolis, Minn. 131 p.

Dennis, R.W.G. 1968. British Ascomycetes. J. Cramer, Lehre, W. Germany. 488 p.

Ellis, M.B. 1971. Dematiaceous Hyphomycetes. Commonw. Mycol. Inst., Kew, Surrey. 608 p.

Fischer, G.W. 1953. Manual of the North American smut fungi. Ronald Press, New York. 343 p.

Hess, W.M., M.M.A. Sassen, and C.C. Remsen. 1968. Surface characteristics of *Penicillium* conidia. Mycologia 60:290–303.

Horak, E. 1968. Synopsis generum Agaricalium [in German]. Beitr. Kryptogamenflora Schweiz (Switzerland) Vol. 13. 741 p.

Ingold, C.T. 1971. Fungal spores: their liberation and dispersal. Clarendon Press, Oxford. 302 p.

Kramer, C.L., and S.M. Pady. 1960*a*. Kansas aeromycology IX: Ascomycetes. Trans. Kans. Acad. Sci. 63:53–60.

Kramer, C.L., and S.M. Pady. 1960*b*. Kansas aeromycology XI: Fungi Imperfecti. Trans. Kans. Acad. Sci. 63:228–238.

Kramer, C.L., S.M. Pady, and C.T. Rogerson. 1959*a*. Kansas aeromycology III: Cladosporium, Trans. Kans. Acad. Sci. 62:200–207.

Kramer, C.L., S.M. Pady, and C.T. Rogerson. 1959*b*. Kansas aeromycology IV: Alternaria. Trans. Kans. Acad. Sci. 62:252–256.

Kramer, C.L., S.M. Pady, and C.T. Rogerson. 1960*a*. Kansas aeromycology VIII: Phycomycetes. Trans. Kans. Acad. Sci. 63:19–23.

Kramer, C.L., S.M. Pady, and C.T. Rogerson. 1960*b*. Kansas aeromycology V: Penicillium and Aspergillus. Mycologia 52:545–551.

Kramer, C.L., S.M. Pady, C.T. Rogerson, and L.G. Ouye. 1959. Kansas aeromycology II: Materials, methods and general results. Trans. Kans. Acad. Sci. 62:184–199.

Lodder, J., ed. 1970. The yeasts: a toxonomic study. 2nd ed. North-Holland Publ. Co., Amsterdam. 1385 p.

Martin, G.W., C.J. Alexopoulos. 1969. The Myxomycetes. Univ. Iowa Press, Iowa City. 561 p.

Pady, S.M., and C.L. Kramer. 1960*a*. Kansas aeromycology VII: Smuts. Phytopathology 50:332–334.

Pady, S.M., and C.L. Kramer. 1960*b*. Kansas aeromycology VI: Hyphal fragments. Mycologia 52:681–687.

Pady, S.M., and C.L. Kramer. 1960*c*. Kansas aeromycology X: Basidiomycetes. Trans. Kans. Acad. Sci. 63:125–134.

Pegler, D.N., and T.W.K. Young. 1971. Basidiospore morphology in the Agaricales. Beih. Nova Hedwigia 35. 210 p.

Raper, K.B., and D.I. Fennel. 1965. The genus Aspergillus. Williams & Wilkins Co., Baltimore, Md. 686 p.

Raper, K.B., and C. Thom. 1949. A manual of the Penicillia. Williams & Wilkins Co., Baltimore, Md. 875 p.

Rogerson, C.T. 1958. Kansas aeromycology I: Comparison of media. Trans. Kans. Acad. Sci. 61:155–162.

Solomon, W.R. 1967. Aeroallergens. Pages 398–436 *in* J.M. Sheldon, R.G. Lovell, and K.P. Mathews. A manual of clinical allergy. 2nd ed. W.B. Saunders Co., Philadelphia.

Toussoun, T.A., and P.E. Nelson. 1968. A pictorial guide to the identification of *Fusarium* species according to the taxonomic system of Snyder and Hansen. Pa. State Univ. Press, University Park. 51 p.

Vries, G.A. de. 1952. Contribution to the knowledge of the genus *Cladosporium* Link ex Fr. Bibl. Mycologica 3. (Reprinted 1967 by J. Cramer, Lehre, W. Germany). 121 p.

Waksman, S.A. 1961. The Actinomycetes. Vol 2. Classification, identification and description of genera and species. Williams & Wilkins Co., Baltimore, Md. 363 p.

Zycha, H., R. Siepmann, and G. Linnemann. 1969. Mucorales [in German]. J. Cramer, Lehre, W. Germany. 355 p.

11

INTERPRETING AND REPORTING THE DATA

Even though pollen concentration measurements are taken with a well-exposed sampler of an acceptable type having a known efficiency for the pollen sampled, the data obtained must be evaluated with an understanding of the many variables which may have influenced the measurement and must be reported in a form useful to the user of the data. A sample may be an accurate measure of the average concentration that existed at *that* location during *that* time period; but it may not be representative of concentrations over a longer or shorter time period, over a wider area, at a different location nearby, or at a different altitude at the same location. Pollen concentrations depend on many variables including distance and direction from sources, height above the ground, season, time of day, weather conditions, terrain, and nearby obstacles to free air flow. Thus, some judgment and experience may be helpful in assessing the representativeness of either a single sample or a series of measurements.

Methods of recording and reporting depend upon the type of data obtained, the purpose for which it is taken, and (to some extent) on individual preference or custom. However, adoption of standard reporting procedures would facilitate comparison of data between stations and should be encouraged. It is particularly important that adequate background information accompany the pollen measurement. Examples are: location and height of measurement, type of sampler, date, time, and some description of weather conditions. Equipment malfunction, accidental modification of a sample, or any unusual event that might affect the validity of the measurement should be reported.

Pollen measurements may be taken for many reasons.

1. Daily measurements reported to the news media for the guidance of allergists, allergic individuals, and the interested public.
2. Accumulation of data over a period of years to serve as a guide for predictive purposes.
3. Surveys to determine seasonal changes in quantities and types of airborne pollens present, the magnitude of their concentrations, or their geographic distribution.
4. Other research studies.

POLLEN COUNTS FOR PUBLIC DISSEMINATION

During the ragweed pollination season, daily pollen measurements are commonly taken in many cities by a hospital, public health agency, or individual allergist and the results made available to the news media. These pollen counts are usually for a 24-hour period from some convenient daylight hour to the same time the next day. They refer to ragweed pollen only and, even today, are mostly obtained with the Durham sampler. For many years, the reports attempted to indicate the average number of ragweed pollen grains per cubic yard of air during the 24-hour period. This value was obtained by use of a conversion factor later shown to have little validity. After it was realized that counts from this sampler could not be reliably converted to number per unit volume, the "count" has indicated only the average number of grains deposited on a one square centimeter surface of the microscope slide. The deficiencies of such measurements were pointed out in Chapters 5 and 6.

A more meaningful daily pollen count is the average number of grains per unit volume and such counts must be obtained through use of samplers whose sampling rate and efficiency are known. These data are generally reported as the average number per cubic meter of air. Although it is, perhaps, most desirable to report the average concentration for a calendar day (midnight to midnight), this is seldom convenient for the changing of samples unless automatic equipment is available.

Reporting the average pollen concentration for a 24-hour period gives no information on the variation in concentration during this period even though the peak concentration during a day may have more relationship to patient symptoms than the average concentration. If such information is desired, sequential sampling (the taking of a consecutive series of samples) must be employed.

For some purposes, it might be more useful to report the concentration for a short, very recent, period of time to better describe the current situation than to report yesterday's average. This would be considered a spot sample. For example, it would be possible, a few minutes after ten o'clock, to report the average concentration between nine and ten. In reporting such a count, however, there should be no intimation that this hour's average count may indicate the day's average count.

To be meaningful, the pollen count reported via newspapers, radio, or other means should be accompanied by a statement that clearly indicates whether it is a deposition or a volumetric count, the location where the sample was obtained, and the time period involved. Lack of such data results in useless and misleading information.

AVERAGE ANNUAL OR SEASONAL CURVES

Pollen counts reflect past conditions even if disseminated promptly. Reasonably accurate predicted counts for a day or so ahead would almost certainly be more useful to those with a need for such information. So far, the few preliminary attempts to forecast pollen concentrations have given hopeful but not conclusive results. Moreover, no known public or research agencies are presently staffed or funded for operational day to day forecasts. However, it is anticipated by some workers in the field that such forecasts will eventually prove feasible and have an accuracy comparable to that of weather forecasts. To serve as a basis for such predictions, data on daily ragweed pollen concentrations must be taken over a period of years, in each region of interest, using a sampler giving an accurate volumetric measurement.

Measurements taken in a single year will show great day to day fluctuations, although some indication of a seasonal pattern is usually apparent. If counts taken on the same date over a period of years are averaged and plotted as a function of time, a smoother curve results. Such a curve, showing the average change in pollen concentration, and additional curves drawn through the higher and lower points, give a good indication of the most likely concentration to be expected and the probable range about this mean. A graph of this nature, utilizing nine years of data, is shown in figure 11-1. If the deviations from the mean are related to weather conditions or other variables through adequate records, the basis of a forecasting system has been established.

POLLEN SURVEYS

Pollen surveys are often taken to determine the types of pollens, and sometimes other biological particles, present through the year and to obtain some measure of their actual or relative abundance. Surveys may also be taken to measure the variation in average concentration of a single type, such as ragweed pollen, in a

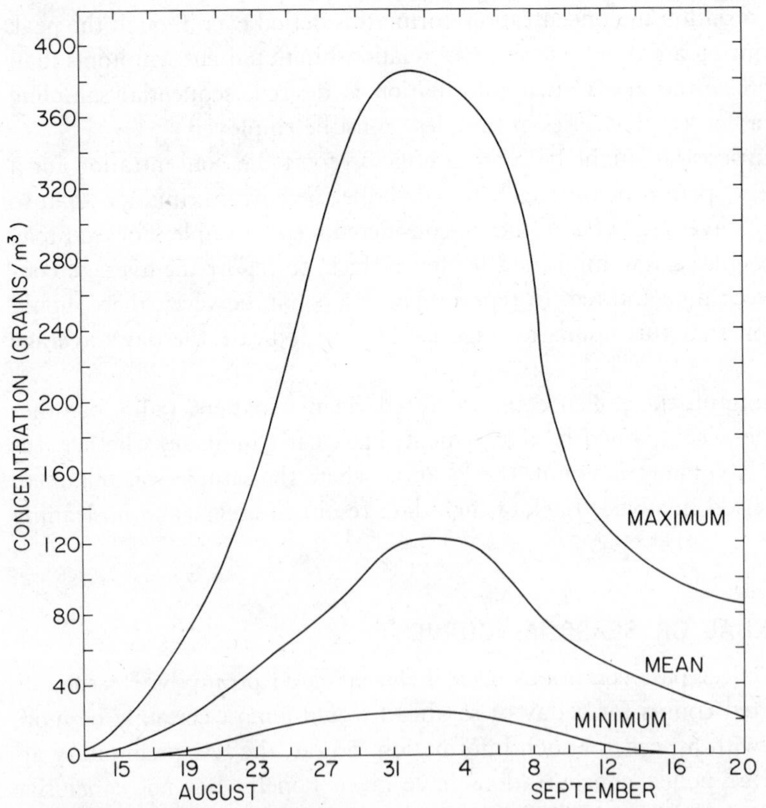

FIGURE 11-1 Mean annual curve of maximum, minimum, and average ragweed (Ambrosia) pollen concentrations at Brookhaven National Laboratory, 1961 to 1969.

city or, by use of multiple sampling stations, over a wider geographical area. At least two studies combined both types of investigation in statewide surveys of all major groups of airborne pollens and fungus spores (Hyland et al. 1953, Ogden and Lewis 1960). However, both were conducted with the Durham sampler, and repetitions using modern sampling equipment would be useful.

Survey data are obtained only at the expense of considerable time and effort. In addition to meeting the objectives of the individual or organization taking the survey, the data may have wider usefulness, particularly for comparative purposes. Comparisons may be made with concentrations at other localities over the same time period or with conditions of the same station in earlier or later years to document longer-term trends. For these reasons, survey data should be taken with the best available sampling device, should be tabulated in a manner usable by others, should be made available to other investigators.

Pollen survey reports vary from simple typewritten sheets to articles in printed bulletins or scientific journals. The report should include the location of sampling stations, the duration of the surveys; the type, time cycle, elevation and exposure of the samplers; the length of the sampling period, and the species or

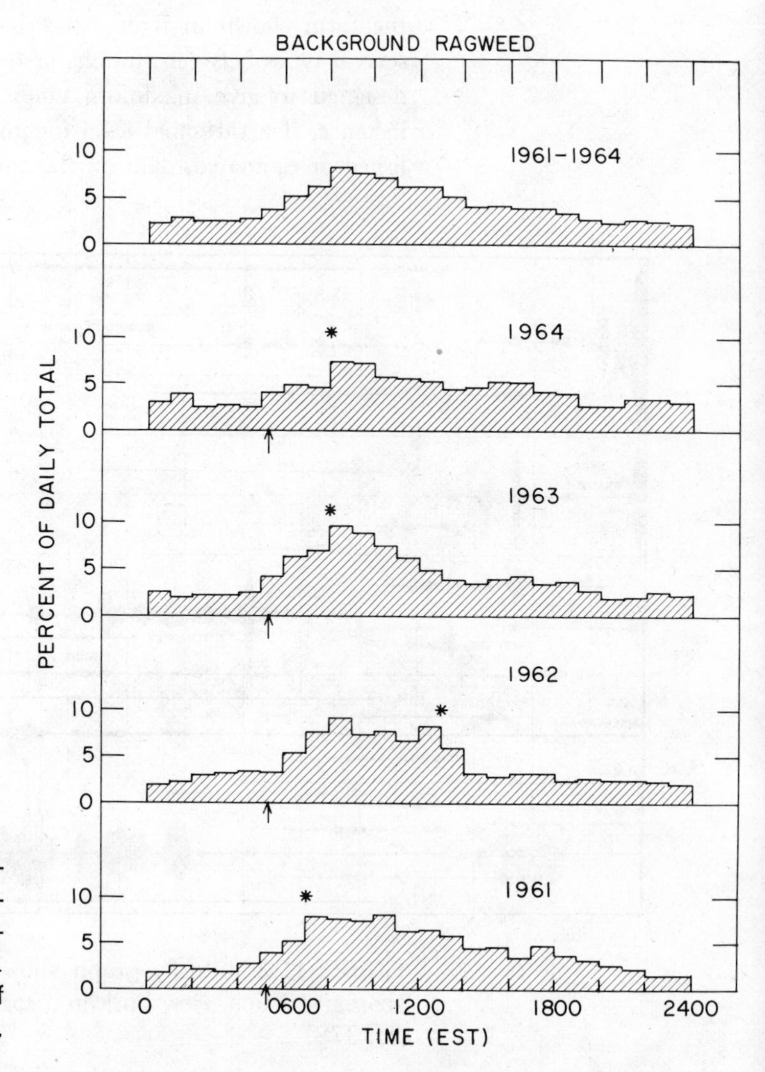

FIGURE 11-2 Histograms showing diurnal variation in ragweed (Ambrosia) pollen concentrations from background sources at Brookhaven National Laboratory in the years 1961 to 1964. Arrows mark the mean time of sunrise; asterisks the time of sample change.

groups of organisms for which the samples were examined. Data may be qualitative or quantitative, but the latter is preferred. Pollens (and other biological organisms) should always be referred to by their scientific names. Generally, the genus name alone is sufficient, although in many cases, pollens can be identified only to family or to groups that are morphologically similar. The name used should not imply greater accuracy than is attained. Commonly used or carefully chosen common names may be useful to some readers and may be included but never to the exclusion of the scientific name.

The survey report should indicate seasonal occurrence with average concentrations for chosen periods of time. These time periods may be daily, weekly, monthly, or even seasonal, depending on the probable use of the report and the page space available. Although the basic data are usually presented in tabular fashion, graphs of several types are excellent visual aids and facilitate comprehension of the data. Perhaps the simplest types to prepare are the line graph (Fig. 11-1) and the histogram or bar graph (Fig. 11-2). The butterfly graph (Fig. 11-3) has the advantage of including a great deal of information in a limited space but is more time-consuming to prepare.

In order to encourage uniform reporting of seasonal data in a meaningful way, the form shown in figure 11-4 is suggested. The data permit comparisons of several types between stations or between years at the same station. The form is designed to give maximum guidance to allergists and pollinosis patients. For instance, if a threshold level for initiation of allergic symptoms should be established or estimated, data on this form will indicate not only the number of days

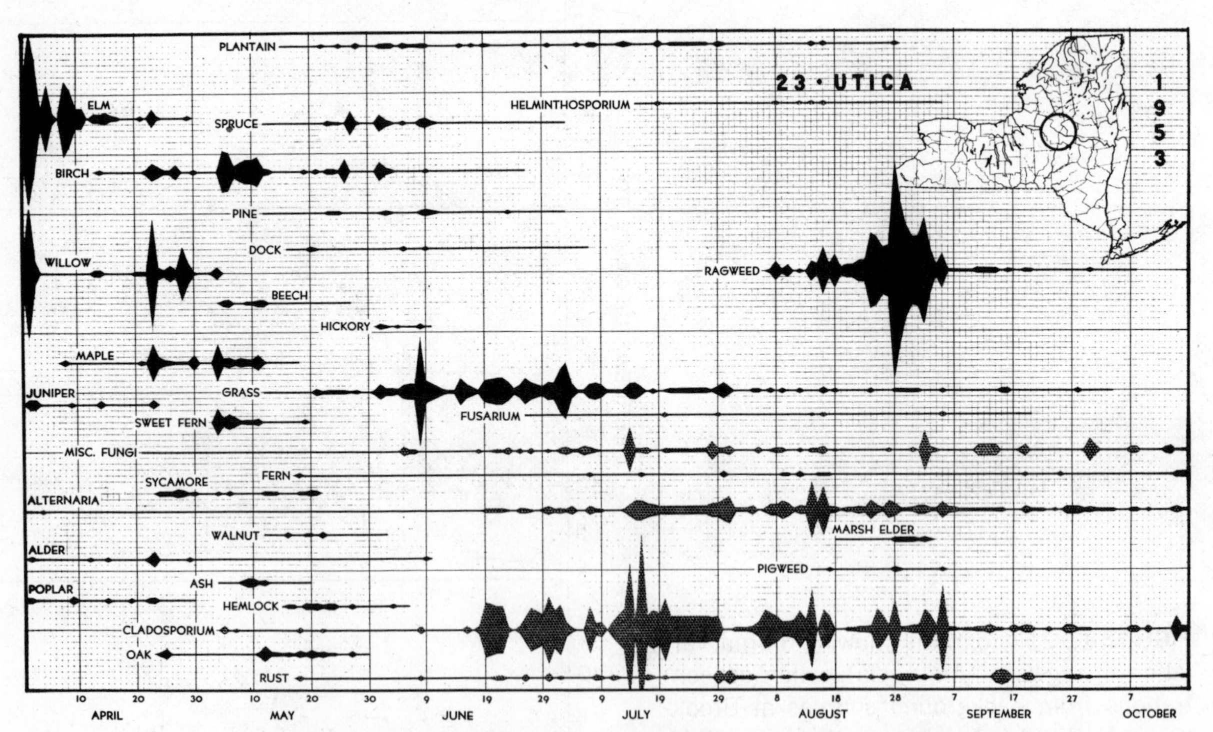

FIGURE 11-3 Butterfly graph showing occurrence of pollen grains and fungus spores at Utica, New York in 1953. From N.Y. State Mus. and Sci. Serv. Bull., No. 378.

SEASONAL SUMMARY

Station____________________

Operator(s)____________________

Location____________________ Height above ground__________

Sampler____________________ Units__________

Sampling Period: Year ________ Date: First__________ Last__________

Pollen____________________

Concentrations: Mean Daily__________ Maximum Daily__________

Frequency of Concentrations Equal to or Above (grains/m^3)

Having this count or greater	1	10	25	50	100	250	500
No. of days							
First date							
Last date							
Length of period (days)							

FIGURE 11-4 Suggested form for seasonal summary reports of pollen concentrations.

above this level but the length of the season and the inclusive dates during which such days occur. This duration may be quite different at two locations having the same mean daily concentration or even the same number of days above this specified level.

RESEARCH STUDIES

Many research programs include pollen sampling but their purposes, methods, and applications are so diverse that no standard method of data handling can be recommended. Usually, the nature of the study determines the analytical methods, although workers in various disciplines have their own established procedures. Some studies will require statistical analyses. Guidance can be obtained from any of the many available texts on statistics which range from elementary to advanced.

Information on graphical methods, beyond the elementary forms mentioned above, is less readily available, but texts should be available in larger libraries.

A brief consideration of several past, present, and potential scientific studies involving pollen sampling is given below. The list covers a wide scope of activities in a variety of fields. It may suggest paths for additional research but is presented primarily to convey some appreciation of the varied relationships which must be interpreted and presented.

1. Clinical studies. These include the relationship between symptoms in allergic patients, as evaluated by a qualified physician, and measurements of pollen concentrations in either a natural or controlled environment. Such studies involve correlation of medical with pollen sampling data. Sampling may also be used to determine which allergic particles are present at the onset of symptoms as an aid in discovering the specific aeroallergen to which the patient is sensitive.
2. The effectiveness of pollen removal devices. The effectiveness of air conditioners, electrostatic precipitators, and other air cleaning devices may be evaluated by sampling before and after use of the device, or by sampling in identical rooms with and without the apparatus. Results may be related to particle size or other variables.
3. The effectiveness of structures in reducing pollen concentrations. Sampling may be conducted inside typical rooms in homes, schools, etc., and concentrations compared to those measured simultaneously outside. Ventilation of the room and weather conditions would be important variables to which the measurements might be related.
4. Ragweed eradication programs. The effectiveness of attempted ragweed eradication programs can be properly evaluated only by several years of sampling both before and after local eradication, since year to year variability in ambient pollen concentrations is large. Results might be obscured by weather effects if conditions in the pre- and post-eradication years are not similar.
5. Dispersion studies. These involve sampling at several distances downwind of a known source, and sometimes at several heights, to determine the rate of decrease in concentration with distance. Since this rate varies with source size and shape, with the nature of the terrain traversed by the particles, and with meteorological conditions, the data must be related to these variables.
6. Pollination effectiveness. In agriculture, the effectiveness of the pollination process may depend on weather conditions during the pollination season. Such studies may involve measurements of pollen emission and measurements of fruit or grain production evaluated in terms of pertinent meteorological and biological factors.
7. Determination of isolation distances. In agriculture, production of pure or hybrid seed of wind-pollinated species depends on isolation of the field from crops of the same species whose pollen might contaminate the particular strain being grown. The necessary separation distance may be determined by dispersion studies of the type described above.
8. Geochronology. Interpretation of fossil pollen profiles from lake or bog sedi-

ments may be facilitated by studies of dispersion and deposition of modern pollens of the same or similar species.

9. Diurnal emission patterns. Diurnal patterns of pollen emission may be determined by sequential sampling in, or downwind of, a stand of plants. Since emission is affected by weather conditions, the data must be interpreted in terms of simultaneous meteorological measurements.
10. Sampler calibration and comparison. Such studies involve comparisons between an experimental sampler and one of known efficiency. Results may be a function of wind speed or other variables and are usually expressed as an efficiency in percent.

References

American Academy of Allergy. Statistical report of the Pollen and Mold Committee. Prepared and published annually for the Committee by Ross Laboratories, Columbus, Ohio.

Hyland, F., B.F. Graham, Jr., F.H. Steinmetz, and M.A. Vickers. 1953. Maine airborne pollen and fungous spore survey. Univ. of Maine, Orono, Maine, 97 p.

Ogden, E.C., and D.M. Lewis. 1960. Airborne pollen and fungus spores of New York State. N.Y. State Mus. and Sci. Serv. Bull. 378. 104 p.

Raynor, G.S., and J.V. Hayes. 1970. Experimental prediction of daily ragweed [pollen] concentration. Ann. Allergy 28:580–585.

Glossary

The definitions in this glossary are for usage within this Manual. Some of the terms also have a broader meaning, or a different meaning in some other publications.

acuminate: tapering to a long, slender point

aerobiology: study of the passive biological constituents in the atmosphere.

aerodynamic: of or pertaining to the air. Used to describe forces caused by atmospheric motions, particularly as these act upon or are modified by obstacles in the air stream or act upon bodies moving through the air.

aeromycology: the study of airborne fungi.

aeropalynology: the study of atmospheric pollens and other plant spores, primarily their identification and dispersal.

air-spora: the population of small airborne particles of plant or animal origin.

allantoid: sausage-shaped, slightly curved with rounded ends.

allergen: environmental agent which causes a disease of allergy. A substance which has the ability to produce an allergic reaction.

allergenic: causing allergic sensitization.

allergy: a state of hypersensitivity to certain things, such as pollen.

ambient air: air which is unconfined and not subject to artificial restrictions on its motion.

anemometer: a device for measuring wind speed.

anemophilous: pollinated by wind.

angiosperm: a plant which bears seeds enclosed in a fruit.

anther: that portion of the stamen in which the pollen grains are formed.

anthesis: the time of flowering or period of pollination.

aperture: see germinal aperture.

apiculus (pl. *apiculi*): a short projection of the cell wall at one or both ends of a spore.

arcus (pl. *arci*): a curving band of thickened exine connecting adjacent pores of some pollens. *Alnus* is a good example.

ascospore: the sexual spore of an ascomycete borne in the saclike ascus.

ascus (pl. *asci*): the saclike cell in which the ascospores are produced. See figure 10-6A.

asexual spore: a spore having a nucleus with the same number of chromosomes as the structure from which it was formed.

aspidate: having shield-shaped thickenings (aspides) surrounding the pores of some pollens.

basidiospore: the sexual spore of the Basidiomycetes. It is borne on a basidium.

basidium: the organ in Basidiomycetes which bears the basidiospores. See figure 10-4A.

betuloid: referring to pollens resembling those of *Betula.*

bilateral symmetry: with two vertical planes of symmetry. Compare with radial symmetry.

bladder: see wing.

body: in winged pollen, the major or central portion bracketed by the wings. It includes a proximal (cap) and distal (furrow) portion.

botany: the study of plants. Plant science.

callose: a common carbohydrate constituent of plant cell walls acting as a rather effective diffusion barrier.

cap: the thick-walled, proximal portion of the body of the winged pollen of conifers. Also called cappus.

centistoke: one hundredth of a stoke. A stoke is a unit of measurement for viscosity.

clamp connection: a hooked, hyphal outgrowth which may occur at cross walls of hyphae of basidiomycetes.

class: a taxonomic category ranking above an order. Each class is composed of one or more orders.

colpate: pollen with germinal furrows only.

colporate: pollen with furrows and secondary apertures (pores or furrows).

compound grain: nature pollen which does not separate, remaining in multiples of four cells.

conidium: an asexual spore, usually of the class Deuteromycetes.

conifer: cone-bearing tree or shrub. The seeds are borne on cone scales.

contaminant: material other than that being sampled or studied.

cyperaceous: pertaining to plants in the Cyperaceae (sedge family).

dehiscence: splitting open. Opening at maturity which releases pollen from anthers, fungus spores from sporangia, seeds from fruits, etc.

dehiscence zone: an area of the wall of some pollen grains which is so modified that it aids in the germination of pollen.

density: mass of any substance per unit volume at a definite temperature.

deposition: settling of small particles to the earth or other horizontal surfaces by gravitational or other forces.

Deuteromycetes: an artificial class of fungi composed of the asexual stages of Ascomycetes and fungi in which no sexual stage is known. Also called Fungi Imperfecti or imperfects. Some authors use the synonymous term Hyphomycetes.

dicot: one of the two great divisions of the flowering plants, having two cotyledons (seed leaves) in the embryo, leaves mostly net-veined, and flower parts usually in fives. Dicotyledonae.

dicotyledonous: refers to dicots.

dioecious: having male and female sex organs on separate individuals; such as pollen produced by one plant, seeds by another.

dispersion: spreading or distribution from a source. Dissemination or scattering.

distal pole: see proximal pole.

distal wall: the pollen wall opposite the center of the tetrad.

diurnal: as used here, the phenomenon of fluctuation in occurence during a 24-hour day.

divergence: a spreading apart of the streamlines of flow as when air approaches an obstacle.

dwarf: much smaller than normal. Commonly used to describe such pollen grains.

ecology: the study of plants and animals in relation to their environments.

efficiency: as used here, the ratio of the number of particles captured by a sampler to the number in the volume of air sampled, before that air was disturbed by the presence of the sampler. The ratio is often expressed as a percentage. Total efficiency may be a function of impaction efficiency (for an impaction sampler), entrance efficiency (for a suction type sampler), and retention efficiency.

electrostatic precipitation: a sampling process in which small, airborne particles having electric charges are subjected to an electric field within the sampler and attracted to an electrode of opposite charge where they deposit. The charge of the particles may be natural or may be imposed as they enter the sampler.

emission: discharging or throwing off, as the release of pollen from the source plants.

endogenous: forming inside another organ of a plant.

endospore: a spore borne inside another cell.

entomophilous: pollination primarily by insects.

equator: the imaginary line encircling the pollen grain midway between the poles.

exine: the outer wall of the pollen grain, often distinctively sculptured and resistant to chemical attack.

family: the taxonomic division between genus and order. Related genera are placed in a family and related families in an order.

fenestrate: pollen sculpture having thin areas giving a window appearance. Used to describe some pollens of the family

Compositae in which ornate spiny ridges enclose open areas.

filament: the stalk of a stamen that supports the anther.

flagellum (pl. *flagella*): a long threadlike appendage. A whiplike locomotory organ.

flora: the plants of a particular area or a publication which contains a listing of them.

fluorescence: emission of radiation, especially visible light, as a result of absorption of radiation from some other source. The glow emitted by some substances when irradiated by ultraviolet light.

foot layer: a thin layer of the pollen grain exine on which the rod layer is based. It may be absent or not recognizable with ordinary light microscopy.

furrow: a structurally reduced elongate area of the pollen wall allowing the exit of the pollen tube. Generally applied to features at least twice as long as wide.

gamete: a nucleus or cell capable of fusing with another nucleus or cell to form a zygote, which in the seed plants becomes an embryo. A germinated pollen grain produces two male gametes, either of which may fuse with a female gamete in the ovary.

genetics: the science of heredity.

genus (pl. *genera*): The taxonomic division between species and family. Related species are placed in a genus and related genera in a family.

geochronology: study of the sequential order of past events as recorded in the earth. These events may be indicated by fossil pollen preserved in sediments or ancient rocks.

germinal aperture: structurally modified, generally reduced area of the pollen wall which aids in the emergence of the pollen tube. It commonly has the form of a furrow or pore which may regulate changes in volume of the pollen grain.

giant: a pollen grain much larger than normal.

granular: used here to describe pollen grain sculpture having elements not much greater than one micron and generally not grouped in an organized manner.

gymnosperm: a seed plant whose seeds are borne on open structures such as cone scales. All conifers are gymnosperms.

helical: used here to describe fungus spores that are spiral or coiled.

hemisphere: the part of the pollen grain from the equator to the pole and often designated as proximal or distal hemisphere.

herbarium (pl. *herbaria*): a collection of preserved plant material, usually systematically arranged. Also, the building where the collection is kept.

hilum: a mark or scar at the point of attachment between a spore and the spore-bearing structure. It is reserved here for basidiospores.

hyaline: colorless.

hypha (pl. *hyphae*): the tubelike cells which compose the vegetative body of most fungi. Collectively known as the mycelium.

impaction: collision by inertial forces of a small, airborne particle with an obstacle or surface in the air stream, usually at right angles to the mean direction of flow.

impingement: collision by turbulent atmospheric motions of a small, airborne particle with a surface, usually not at right angles to the mean direction of flow. Liquid impingement results when the air stream is diverted into a liquid which retains the particles as the air bubbles to the surface.

inaperturate: without an aperture.

inertia: the tendency of a body at rest to remain at rest or of a body in motion to continue in uniform motion in the same direction, unless acted upon by an external force. Used here for the tendency of a small airborne particle to continue in uniform motion.

intine: the inner wall of the pollen grain. Generally of homogenous structure and rather easily degraded chemically.

isokinetic sampling: sampling with a device which does not disturb the speed or direction of air flow approaching the sampler or change its particulate concentration or size distribution in any way.

lapse rate: the rate at which temperature changes with altitude.

lateral view: used here to describe the equatorial view of the irregularly shaped winged conifer pollen.

leptoma: thin-walled area of a pollen grain.

lozenge-shaped: used here to describe a pollen grain with four equal curved sides, but with unequal diagonals.

marginal crest: the juncture of the wings and cap of some conifer pollens may be marked by a thickened projection of the cap. This is the marginal crest or marginal ridge.

meiosis: division of a nucleus in which each of the resultant nuclei have half as many chromosomes as the mother nucleus.

melanized: containing a dark pigment which may be melanin.

meridional: refers to pollen features oriented in the same direction as the polar axis. Meridional furrows are directed poleward as would be lines of longitude on a globe of the earth.

meteorology: the science treating of the atmosphere and its phenomena.

micrometer: an instrument for measuring minute distances. Micrometer is sometimes used as being synonymous with micron.

micron: one-thousandth of a millimeter. Represented by the symbols 10^{-6} m, μ, or μm.

microspore: as used here, the cell that develops into the pollen grain.

molecular membrane filter: a thin, porous filter composed of inert, cellulose esters or similar polymeric materials. All particles larger than the pore size are retained on the surface; also many smaller particles are captured by electrostatic attraction.

monocot: one of the two great divisions of the flowering plants; having one cotyledon (seed leaf) in the embryo, leaves mostly parallel-veined, and flower parts usually in threes. Monocotyledonae.

monocotyledonous: refers to monocots.

monoecious: having male and female parts borne in separate cones or flowers, but with both sexes on the same plant.

morphology: the study of form and structure.

mph: miles per hour.

mps: meters per second. One mps = about 2.24 miles per hour.

mycelium: the vegetative body of most fungi which is composed of tubular elements known as hyphae.

netted: refers to pollen grain sculpture, with elongate elements which interconnect forming netlike meshes.

nomenclature: the system of naming things.

Nuclepore filter: a filter of thin, transparent, polycarbonate film having pores produced by charged particles in a nuclear reactor followed by chemical etching. Its sampling characteristics are similar to those of the molecular membrane filters, but it has greater tensile strength.

oncus (pl. *onci*): a thickening of the intine beneath the pores of aspidate pollen grains.

operculum (pl. *opercula*): a clearly defined thickening of the pore membrane of pollen.

paleobotany: the study of fossil remains of plants.

palynology: the study of pollens, spores, and other tiny organic particles; primarily their identification and occurrence. This includes aeropalynology (airborne particles), paleopalynology (fossil particles), mellitopalynology (honey pollens), etc.

palynomorph: a term applied by paleontologists to include pollen and other plant spores and also other associated biological particles primarily of fungal, algal, and zoological origin. The term might also be applied to similar particles encountered in aeropalynological studies.

papilla (pl. *papillae*): a protuberant area of the pollen or spore wall.

parameter: a quantity in a mathematical equation to which the user can assign an arbitrary value as distinguished from a variable which can assume only those values that the form of the equation makes possible.

phenology: the study of periodic phenomena. An example is the time of flowering in relation to climate.

phialide: a short, flask-shaped stalk from the apex of which spores (phialospores) are produced. Found only in one group of Deuteromycetes.

pistillate: refers to flowers having pistils but no stamens. Unisexual female flowers.

polar: pollen features most closely associated with the poles rather than other portions of the grains (e.g., polar area).

polar axis: an imaginary line connecting the proximal and distal poles.

polar view: the view of a pollen grain in which the line of sight is parallel to the polar axis.

pollen: mature microspores or microgametophytes produced in the sporangia of male cones or the anthers of flowers.

pollen tube: a tubular growth from the germinating pollen grain.

pollination: transfer of pollen from where it was produced to a receptive surface associated with the production of female gametes.

pollinosis: allergy caused by pollen in the respiratory tract. Commonly known as hayfever.

pollutant: as used here, any material, gaseous, liquid, or solid, which is not a normal constituent of the atmosphere, particularly material detrimental to human health, animal or plant life, or aesthetics.

polygamous: having unisexual and bisexual flowers on the same plant.

porate: pollen with germinal pores only.

pore: a structurally reduced, more-or-less circular area of the pollen wall which allows the exit of the pollen tube. Thin or clear area of a fungus spore wall.

proximal: that part of the pollen grain wall which is directed toward the center of the tetrad.

proximal pole: in symmetrical pollen tetrads, lines radiating from the center to the remote outer faces bisect the inner faces of individual grains at the proximal pole, the outer faces at the distal pole.

radial symmetry: having more than two vertical planes of symmetry.

ragweed: any of the species of the genus *Ambrosia.*

re-entrant angle: the angle formed by the juncture of the wings and cap in some conifer pollens, as seen in lateral view.

reticulate: elongate elements interconnecting in a netlike system. A type of pollen or spore wall sculpture.

rpm: revolutions per minute.

rodlike elements: structures found in the pollen wall appearing as pillars; generally forming a middle layer of the pollen exine, supporting the roof, and usually based on the foot layer.

roof: the outer, more-or-less continuous layer of the pollen exine. Other sculpturing elements may be imposed on its surface. It is supported by the rod layer.

saccus (pl. *sacci*): see wing.

sculpture: used here to include all elements of pollen wall construction visible by ordinary light microscopy. This would include surface ornamentation and internal wall details. The latter features are technically described as structure in textbooks of pollen morphology and those sources should be consulted for an understanding of this distinction.

septate: with cross walls.

sequential: consecutive or following in time in an orderly manner. Used for sampling devices which take a number of discrete samples at predetermined intervals.

sexual spore: a spore having a nucleus with half the number of chromosomes as the structure from which it was formed.

smooth: pollen wall sculpture having elements less than one micron in greatest dimension and without organized patterning. Fungus spores described as smooth are completely unornamented.

species: a taxonomic category below genus. A group of closely related interfertile individuals.

spheroidal: resembling a sphere in shape.

spine: any distinct, generally pointed process of a pollen grain or fungus spore wall.

spinule: a small spine.

spiny: with generally pointed processes distinctly larger than granules. A type of pollen wall sculpture.

sporangiospore: spore borne in a sporangium.

sporangium (pl. *sporangia*): a structure containing spores. In fungi, the term is restricted to structures containing asexual spores, especially in Zygomycetes and Oomycetes.

spore: a general term used in different senses. Mostly used for the simple reproductive bodies of plants that do not produce seeds.

sporogenous: as used here, tissue in the anthers which give rise to pollen mother cells.

sporopollenin: a chemically complex substance forming the exine of the pollen wall.

spot sample: sample obtained over a short period of time.

stamen: the pollen-bearing organ of the flower. It is composed of anther and filament.

staminate: having male but not female reproductive organs. Staminate cones or flowers produce pollen but no seeds.

sterigma (pl. *sterigmata*): the pointed outgrowth from the basidium which bears the spore in Basidiomycetes.

striate: streaked by more-or-less parallel sculpturing elements. These may be straight or curved. The striae of pollen walls are generally considered to be the depressions between the higher elevations of such grains.

structure: see sculpture.

tapetum: a tissue, generally one-cell thick, encasing the sporogenous tissue of the anther and providing nutrition for the pollen mother cells and developing pollen grains.

taxon (pl. *taxa*): a general term applied to any taxonomic category, such as family, genus, species, etc.

taxonomy: the systematic classification of organisms. It includes phylogeny and nomenclature.

teliospore: the spore in the rusts or smuts from which the basidium is produced. Sometimes called teleutospore.

terminal velocity: a constant rate of speed attained by a falling object, such as a small particle, as a result of the opposing forces of gravity and aerodynamic drag exerted by the air. Also called settling rate.

tetrad: a group of four adhering objects, such as the four pollen grain cells formed from one pollen mother cell. A union of four cells. See figures 9-1, 9-2, and 9-11C.

thermal precipitation: a sampling mechanism in which the faster moving air molecules from a hotter surface strike small particles in the sampler with greater energy than slower moving air molecules from a nearby colder surface, thus driving the particles in the direction of the colder surface where they are collected.

tracer: a substance such as a gas or small particle emitted into the atmosphere for the purpose of determining atmospheric motions.

transport: carrying a particle from one position in space to another.

transverse furrow: a pollen grain furrow whose long axis crosses the long axis of the meridional furrow at right angles.

triradiate streak: a figure composed of three arms radiating equally from a center. Found on the cap of some *Abies* pollen.

turbulence: irregular or random fluctuating motions superimposed on a mean flow.

urediniospore: also called urediospore or uredospore. Asexual spore in the rust life cycle which causes reinfection by the same stage or which develops into the next stage: the teliospore stage.

vacuole: a membrane-bound cavity in the cytoplasm which contains a solution different from the cytoplasm.

velocity: a vector quantity which includes both speed and direction.

vesicle: see wing.

wall stratification: the separate layers comprising the exine of the pollen wall.

wind velocity: a vector quantity describing the speed and direction of the mean wind.

wing: a bladderlike lateral extension of the body of the pollen grain found in some conifers.

zoospore: a spore having one or more flagella.

Index